Controlled-Release Technology

CONTROLLED RELEASE

the
UNIVERSITY
of
GREENWICH

JOHN WILEY & SONS

New York • Chichester • Brisbane • Toronto • Singapore

Library of Congress Cataloging in Publication Data:

Main entry under title:

Controlled-release technology.

"A Wiley-Interscience publication."
Includes bibliographical references and index.
Contents: An overview of controlled-release
technology / S. A. Patwardhan and K. G. Das—Design
parameters / R. L. Collins—Chemical methods of controlled release / S. A. Patwardhan and K. G. Das—[etc.]
1. Controlled–release technology. I. Das, K. G.

TP156.C64C67 1982   688.8   82–11052
ISBN 0–471–08680–0 √

Printed in the United States of America
10 9 8 7 6 5 4 3 2 1

Contributors

N. F. Cardarelli, Environmental Management Laboratory, University of Akron, Akron, Ohio

R. L. Collins, Physics Department, University of Texas at Austin, Austin, Texas

K. G. Das, Regional Research Laboratory, Hyderabad, India

A. F. Kydonieus, Hercon Division, Health-Chem Corporation, New York, New York

S. A. Patwardhan, National Chemical Laboratory, Pune, India

C. M. Radick, Environmental Management Laboratory, University of Akron, Akron, Ohio

R. M. Wilkins, Department of Agricultural Biology, The University of Newcastle upon Tyne, United Kingdom

L. T. Zeoli, Hercon Division, Health-Chem Corporation, New York, New York

G. Zweig, School of Public Health, University of California, Berkeley, California

Preface

Controlled-release technology probably means many things to specialists in many areas of science and technology. It is an emerging technology for the effective, economic, and safe use of any bioactive toxic chemical or plant nutrient. It provides a unique opportunity to achieve a quantum leap in the benefits of drug therapy by controlling the rate of drug delivery to the site of action. The objective is to achieve selectivity, specificity, and accuracy in delivering the optimum dose of the active ingredient to the desired site at the appropriate time, and to obtain maximum activity on the target while producing minimal effect on the nontarget materials. It is a bridge between polymer and pesticide or drug technologies and may be seen as an attempt to simulate nature's processes.

This technology has its roots in the drug industry. It has spread to other areas such as agrochemicals, plant nutrients, veterinary drugs, and flavors. The high degree of current interest is evident from the amount of work in progress. A few books and many papers are available, and international symposia are held annually.

Some controlled-release formulations have become commercial products. Many are undergoing trials and are in the experimental stage. Developmental work is still in its infancy. Environmental aspects determine the scope and utility of design parameters. Evaluation at the required site of action poses practical problems. The question of chronic vs. acute intoxication offers a challenging problem which evades a satisfactory solution.

Many controlled-release technologies, release mechanisms, design parameters, and chronic vs. acute intoxication are dealt with in this book. Methods for determining release rates and the environmental aspects that govern the commercial success of controlled-release systems are included to make the volume complete.

An overview of controlled-release technology and design parameters are presented in the first and second chapters, respectively. Chapters 3 through 5 deal with chemical, physical, and microencapsulation approaches to controlled-release technology. Chapter 6 looks at release rates while the last two chapters deal with the environmental aspects of controlled-release technology and on chronic vs. acute intoxication, respectively.

The authors have attempted to make their contributions comprehensive, authoritative, and up-to-date, presenting the state of the art to the extent possible in this fast-growing field. I am very much thankful to all of them for their hearty cooperation and for their excellent contributions. I am indebted to the Director, National Chemical Laboratory, for his great support and to the publishers for their enthusiastic collaboration in publishing this volume.

K. G. DAS

Hyderabad, India
September, 1982

Contents

Controlled-Release Technology

CHAPTER ONE

An Overview of Controlled-Release Technology

S. A. PATWARDHAN
National Chemical Laboratory, Pune, India

K. G. DAS
Regional Research Laboratory, Hyderabad, India

CONTENTS

1. INTRODUCTION

The concept of controlled release is a novel approach to the safe and effective use of any toxic active ingredient, whether pesticide, drug, or fertilizer. Controlled-release technology promises to solve a myriad of problems that have in common the application of an active toxic compound to a system in such a way as to accomplish a specific purpose while avoiding certain other possible responses. Ideally, the active agent is released at a controlled rate that maintains its concentration in the system within optimum limits for a desired period. The principal advantage of controlled-release technology is that much less of the active ingredient is required for the same period of activity than is recommended in conventional methods of application. Thus, controlled-release technology holds great promise for improving the efficacy of existing drugs and pesticides and for reducing the problems associated with others.

Many insects and microorganisms are known to cause great damage to grains, vegetables, fruits, wood, paper, cotton, wool, rubber, plastics, and leather. In agriculture, crops have often been damaged or destroyed by insects, rodents, and disease. Rodents inflict tremendous damage on crops both in the field and during storage. Insects, mites, mollusks, nematodes, fungi, bacteria, and viruses are responsible for plant diseases. A large number of phytopathogenic microorganisms and insects infest crops, vegetables, and fruits.

Bats, birds, rats, mites, and parasites are a great danger to animal husbandry. Not only do they cause animal unrest and damage to animal hides, but they are also vectors of infectious disease. The danger to human health from various insect pests that are vectors of infectious disease is difficult to evaluate. Epidemic diseases transmitted by pests include encephalitis, typhus, relapsing fever, sleeping sickness, elephantiasis, and malaria.

Most early efforts to alleviate disease were ineffective because the means of transmission were unknown or little understood. As it became recognized that various pests were not only annoying and materially destructive but also life threatening, more effective methods of pest control were developed and practiced. Both mechanical and chemical methods of pest control have a long history. As early as 350 B.C., Aristotle described the use of arsenic compounds for rodent control. In the 1980s we are engaged in the development of biological methods of pest control.

Pest control has now become a part of daily life. Chemicals are used on the farm, in the garden, and in the house to control both plant and insect pests. Large-scale farming practices of the present century have led to the rapid development of agricultural insecticides. Synthetic pesticides such as chlorinated hydrocarbons, organophosphates, and many systemic insecticides were developed to satisfy special requirements. Some synthetic pyrethrins have been found to be efficient and nontoxic to humans.

Pesticides are toxic chemicals. Many pests have developed resistance to common pesticides that have been in long and continuous use. This increased tolerance has resulted in the substitution of more toxic chemicals for less toxic compounds and in the application of larger doses.

In spite of the remarkable advances in science and technology, we have not found a satisfactory solution to eliminate malnutrition, starvation, and disease and to preserve the environment from contamination. Researchers are busy developing high-yield varieties of crops by tissue culture, hybridization, radiation-induced mutation, and other techniques. Controlled-release technology has a significant role in the integrated approach to pest control. The conflict between the absolute need to use pest control agents in agriculture and public health applications and man's great desire to preserve the environment free from toxic materials is still evading a permanent solution. Controlled-release technology aims at increasing the efficiency of active ingredients and at decreasing their quantity and distribution in nontarget systems and in the environment. Until better control techniques become available, chemical control and the necessity for preserving environmental quality can be reconciled by the successful application of controlled-release technology. Localization, prolongation of the desired action with minimal side effects, and one-time application are some of the unique features of controlled-release systems.

2. PESTICIDES

2.1. Historical

In the seventeenth century nicotine was in use as an agricultural insecticide. Later, carbon disulfide was used as a soil sterilant and Paris green was used to control potato beetles. Chemicals were also applied to the bottom of ships to prevent damage and fouling. Pitch, wax and oil-based preparations containing sulfur and salts of arsenic, lead, mercury, and tin were popular before anti-fouling paints came into existence. Cuprous oxide in shellac or resin base is a known antifouling preparation. Paints containing oxides and salts of mercury, arsenic, and copper were in use in the 1800s. In 1892 the first synthetic organic pesticide, potassium dinitro-o-cresylate, was prepared. Fluorine compounds, pyrethrum, and rotenone made their appearance later on.

DDT was reported in 1939 and it quickly became a universal insecticide. Other chlorinated hydrocarbons, BHC, toxaphene, chloradane, aldrin, and dieldrin were subsequently introduced, followed by organophosphates and other synthetic insecticides.

2.2. Conventional Formulations

Chemicals used to destroy any species of pests are called pesticides. Pesticides include substances that stimulate or retard the growth of plants or repel, attract, and sterilize insects and are classified as acaricides, algicides, arboricides, bactericides, fungicides, herbicides, insecticides, molluscicides, nematocides, and zoocides. Insecticides are divided into subgroups: contact, stomach, and systemic. Fungicides are used to control diseases of growing plants and for seed disinfection before planting. Both selective and nonselective herbicides are known which act by contact or systemic modes.

The high economic efficiency achieved with the use of pesticides in agriculture and other areas has favored the rapid development of the pesticide industry. There has been an increase in pesticide production and a continuous change and improvement in chemical formulations. In addition to their high biocidal activity for various pests, pesticide formulations must be safe to handle during both production and use, nontoxic to humans, domestic animals, useful plants, and beneficial insects and microorganisms. Plants treated with any pesticide must after specified periods contain only such residual amounts that complete safety is assured in their use as food not only for animals but also for man.

In many countries standards have been set for the maximum content of pesticides in foodstuffs for both humans and domestic animals. The intervals between plant treatments are so recommended that at harvest the formulation applied should have completely or nearly completely decomposed so that the food materials do not contain amounts of pesticide residues harmful to human health. In determining the degree of toxicity of a formulation, it is necessary to give attention to its chronic toxicity, the possibility of accumulation in the body, the reversibility of the toxic effect, the route of entry, and a number of other factors, as well as the toxicity of the products of its metabolism.

The success of pesticides depends to a large extent on the formulation and the conditions under which the toxicant is brought into contact with the target. Since both the nature of the target organisms and the chemical structures of toxicants vary widely, it is necessary to produce a large number of formulations suitable for practical applications. The most important types of formulations include granules, wettable powders, solutions in water and organic solvents, emulsive concentrates, and aerosols and fumigants. The selection of the most appropriate, efficient, economical, and safe formulation depends on the physicochemical properties of the active agent, its purpose, and the mode of application. Persistent pesticides leave toxic residues and are incorporated into the food chain. The less persistent pesticides tend to be less effective and call for repeat applications and higher doses. Excessive amounts are usually applied to make up for the losses due to biodegradation, evaporation, and leaching. In actual field applications the quantities are often further increased, since in conventional methods as much as 60–90% never reaches the target organism.

It is desirable to place pest control on a sound quantitative and predictive basis. Conventional pesticide applications are wasteful, since most of the pesticide applied is not used for actual control of the pest but functions as an imperfect reservoir to maintain the critical control level. To achieve satisfactory control for 50 days with a typical nonpersistent pesticide with a half-life of 15 days, the level of conventional application would have to be tenfold the minimum requirement. If the period of control is to be doubled or tripled, the level of application has to be increased 100- or 1000-fold, respectively. Theoretically the addition quantity required to prolong the period of control is only the quantity necessary to replace the fraction lost so that the active ingredient level is maintained for effective pest control.

2.3. Disadvantages of Conventional Formulations

Inorganic compounds used as insecticides and fungicides containing antimony, boron, copper, fluorine, manganese, mercury, selenium, sulfur, thallium, and zinc as active ingredients are known to persist in the soil, and their residues are harmful to crops. Chlorinated hydrocarbons are even more persistent. More than a thousand pesticides are now in common use, a few persist for more than a few weeks or at most months. In the early days there was little anxiety as to possible long-term ecological hazard caused by their use even though there was some evidence that large residues in the soil could be phytotoxic. Small quantities of pesticide residues were reported in plants and animal tissues. There were instances of fish being killed when water was sprayed with antimalarials. Pesticide residues have been detected in the air we breathe, in the streams and lakes that supply drinking water, in the cloth we wear, and in our bodies. Many birds and forest animals have become victims of these poisonous pesticide residues.

Among the environmental contaminants, chemical pesticides occupy a unique position. Their distribution in the environment depends on the way they are applied and on their volatility and solubility. Their persistence in terms of hydrolysis, oxidation, and removal by adsorption is governed by their physical and chemical properties. Biological properties control their utility and toxicity to nontarget living things. The mode of application determines which part of the environment becomes contaminated. Aerial application can contaminate simultaneously, air, soil, and surface water. The fate and persistence of pesticides pose another complex problem. Their translocation, metabolism, and degradation are also complicated.

Toxicological research a decade ago lacked the present conceptual approaches to understand the mechanisms involved in cellular biochemistry relating to cellular alteration and ultimately to disease states. Over the past decade a gradual evolution in the fields of cellular biology, cytogenetics, and biochemistry have focused attention on the significance of subthreshold toxic stresses to induce responses and pathological changes at the cellular level. Studies are in progress on the biological impact of pesticides on humans and their environment. Protracted cellular insults may occur from chronic exposure to pesticides. Safety measures based on short-term observation of single or multiple exposures to a pesticide are not adequate.

Spurred on by the threat of an impending environmental upset and the critical need to feed the rapidly increasing world population, newer and safer methods of pest control are being quickly developed. Biological control holds great promise. In this approach a species harmless to man, crops, and animals but pathogenic to or a predator of a pest species is reared and released to the pest-infected area. Insects sterilized with irradiation or chemicals or with altered genetic characteristics are released to mate with other members of their species and ultimately eliminate the pest population. In another technique to induce sterility in the wild population, sterilizing agents are introduced into the environment. The use of pheromones as attractants for insects is undergoing fast development. Genetic control is another approach to insect control. These control methods must be explored with great caution.

2.4. Controlled Release

In the recent past many persistent pesticides have been phased out because of the environmental and toxicological hazards they posed. Less persistent, but sometimes more acutely toxic, pesticides have replaced them. The newer pesticides often generate other problems such as greater chances for accidental exposure of humans to highly toxic concentrations, and the need for multiple applications because of lower persistence. Controlled-release technology offers an ideal solution to these problems. The insect growth regulator methoprene is so unstable in the aquatic environment that its practical application is possible only with controlled-release methods. Some of the newer pesticides may never reach the market unless they are stabilized long enough to effect control through controlled-release technology. Economic and environmental advantages are also gained by the constant release of lower concentrations of toxicants than are possible with conventional formulations.

In normal practice, relatively high doses of pesticides, drugs, fertilizers and other biocides are administered at periodic intervals. Immediately following an application the concentration of the active ingredient rises to a high level in the system treated. This initially high concentration may produce undesirable local effects in the target area or contaminate the environment. As time passes, the concentration begins to fall because of natural processes such as elimination, consumption, or degradation, and before the next application, it may fall below the optimum level for the desired response. Thus the conventional modes of application are usually rather inefficient, since a considerable quantity of the active agent never performs the desired function. These factors inflate the cost of treatment.

In a controlled-release formulation the active ingredient is released at a continuous and constant rate for a predetermined period so that only the target is attacked. The active agent is localized, and the amount released is just enough to achieve the desired effect. In fact, to maintain the pesticide levels by a controlled-release mechanism, the amount of active material needed for 50, 100, and 150 days of control will be about 3, 6, and 8 mg, respectively, instead of the levels of 10, 100, and 1000 mg that are conventionally recommended and used. Minimal damage is caused to nontarget life forms, and environmental contamination is negligible.

One of the problems in controlled-release technology is to combine the active agent with a degradable carrier in an economic manner and achieve a release profile that best suits the requirement. Many techniques are available for the design and preparation of controlled delivery systems. They include dissolving or physically trapping the active agent in an appropriate natural or synthetic polymeric matrix or chemically binding it to a suitable polymer. Pesticides have also been microencapsulated in natural and synthetic polymers. Natural film-forming materials or microcapsules prepared from gelatin, cellulose derivatives, and synthetic polymeric films have been widely used for microencapsulation. Phase-separation reactions and interfacial polycondensation reactions have been successfully applied for microencapsulation of pesticides. Encapsulation of pesticides within starch xanthate matrix has been reported.

3. DRUGS

3.1. Dosage Forms

Great progress has been made in the management of disease through the introduction of many life-saving drugs. These accomplishments in drug development are not balanced by similar growth of drug delivery systems. A drug will not be beneficial to the patient to the maximum extent unless it is delivered to the target area in the right concentration at the right time. The concentration delivered should be such that side effects are minimized and the therapeutic effects are maximized. If the desired target tissue and the site of administration are well separated, the drug may have to pass through many body barriers.

Over the years a variety of modified drugs and dosage forms have been available to control drug action. Dosage forms include prodrug, controlled release, sustained release, prolonged release, and timed release preparations. In all of these, some degree of control has been obtained over drug placement. A maximization of therapy has not been achieved.

Controlled drug release is the phasing of drug administration so that an optimal amount of drug is made available to cure or control the condition in a minimum time with minimal side effects. Sustained-release and prolonged-release dosage forms prolong drug levels in the body.

3.2. Limitations of Conventional Dosage Forms

The conventional dosage forms, which include tablets, capsules, injectables, and eye drops, share the inability to maintain the drug level in body tissues. The main function of a conventional dosage form is to convey a unit dosage from a container to the body tissues. When a tablet or capsule breaks down to empty its drug content into the stomach, the drug starts dissolving in body fluids and is absorbed into the bloodstream. This leads to the distribution of the drug in the body and to the target tissues. The concentration in the body rises rapidly to a peak, followed by a fall which is determined by its characteristic metabolism and elimination pattern. This cycle is repeated with each dose. With each administration of the drug, its concentration passes through the therapeutic level. Drug concentrations may thus fluctuate between levels that can cause side effects and others that are so low as to be nontherapeutic. This inadequate and inefficient functioning of conventional dosage forms is partly responsible for the toxic side effects of well-established old drugs as well as of the newer ones. Many potent life-saving drugs destined to act only on specific target organs have to cross many body barriers through the bloodstream in high concentrations in order to maintain a therapeutic drug level.

Considerable research has been directed toward a better understanding of the mechanism of drug absorption from the different routes of administration. Studies are also undertaken on the effects of drugs at absorption sites and on the possible injury to subcutaneous and intramuscular sites. Parenteral injection has been recognized as an essential part of therapy. Injectable formulations lead to a local lesion at the site of injection, and satisfactory tolerance requires specific adjustments of solubility, concentration, and volume. An injectable substance with moderate musculoirritative properties can cause a focus of neurosis which is healed by regeneration of striated fibers. Repeated injection of irritating substances results in progressive atrophy of muscle fibers and replacement fibrosis. A number of injections of substances with the lowest irritative index can elicit adaptive lymphatic response. A more serious hazard is nerve injury. Underestimation of the damage caused by repeated injection into the quadriceps femoris muscle has resulted in sequelae of contracture in the upper legs of infants.

3.3 Controlled Release

Over the years, long-acting dosage forms have been developed. Such products are mainly of three types—sustained release, prolonged action, and repeat action dosage forms. These formulations deliver initially the amount required to trigger the desired pharmacological effect. A maintenance dose is supplied at the same rate as the rate of drug removal from the body over the time interval for which the pharmacological response is required, but a constant drug level is not maintained.

In practice it is difficult to design an ideal sustained-release drug dosage formulation, and most dosage formulations are the prolonged-action type. For the development of such formulations there are several stages of activity from inception to completion. The interacting factors relating to the drug, the disease, and the mechanism of prolongation are important to the formulator. The choice of drug is governed by drug properties, the route of drug delivery, whether the therapy is acute or chronic, the nature of the disease, the patient, and the appropriate delivery system. The behavior of the drug in the delivery system and in the body are two important factors that contribute to the success of the system.

An appropriate drug must have the desired therapeutic efficiency, pharmacokinetic behavior, and physicochemical properties that permit effective transfer from the delivery port to the target tissue site. In controlled-release systems, the shorter the half-life of the drug, the more closely will the temporal pattern of its concentration in blood or other body fluids follow the temporal pattern of drug administration. In conventional dosage forms, the rule is the longer the half-life the better. An appropriate drug-delivery module is bioengineered to contain and protect the drug in a reservoir from which it is released at a rate precisely determined by a rate controller. The module also contains an energy source to effect the transfer of drug molecules from the reservoir to the body site. The platform contains the drug-delivery module and ensures correct positioning and safe coupling of the system to the tissue site. The system delivers a drug program whose rate and duration of drug delivery will perform the desired therapeutic function.

Physicochemical properties and biological factors influence the design and performance of sustained-release dosage forms. Among the physicochemical properties, dose size, aqueous solubility, partition coefficient, drug stability, protein binding, and molecular size are important. In most cases, these properties are restrictive rather than prohibitive, making the design of sustained-release systems difficult. The pharmacokinetic properties and biological response parameters have working range for the design of these dosage forms. Absorption, distribution, metabolism, biological half-life, incidence of side effects and margin of safety of the drug have to be examined and evaluated. Some of the limiting factors in developing an effective sustained-release dosage form are extremes of aqueous solubility, oil/water partition coefficient, erratic absorption properties, multicompartment distribution and binding, extensive metabolism/degradation during transit from delivery site to target tissues, and narrow therapeutic index. Disease state and circadian rhythm have to be considered also, even though they are not drug properties. These limitations can be overcome by the application of physical, chemical, and biomedical engineering approaches singly or in combination.

The term controlled drug delivery is used in a rather loose sense. Delivery of a constant tissue drug level has not yet been achieved. All the successful dosage forms on the market today only prolong drug levels; they do not control the delivery. They do not maximize drug utilization and do not take into account changes in the drug requirement during the course of treatment, that result from circadian rhythm, changes in the pathological state, patient variation, and so on. However, these drug delivery systems are a significant step forward compared to their conventional counterparts with respect to temporal drug level control and patient compliance.

One of the difficulties in developing controlled delivery systems is the technological limitation. Dissolution control systems have limited flexibility. Greater flexibility is possible with drugs in polymers. Orally administered sustained-release formulations have a limited duration of action and offer a formidable challenge, owing to the transit time in the hostile environment of the gastrointestinal tract; this time varies from patient to patient. The difficulties encountered with sustained-release dosage forms that prolong drug levels in the blood are surmountable, since the intramuscular route offers a more controlled environment for their application. The physical modification approach has been tried for prolonging action by administering through the parenteral route.

Implants are the most successful sustained drug delivery systems on the market. They are delivery devices containing the drug in a polymeric capsule or matrix. Silicone and high density polyethylene have been widely used in the manufacture of intrauterine progesterone contraceptive systems because of their physiological inertness and biomedical compatibility (Figs. 1 and 2). Hydron, an implantable hydrogel, and contact lenses are prepared from copolymers of polyhydroxyethyl methacrylate and ethylene glycol dimethacrylate. The Ocusert system was the first in a family of therapeutic systems capable of controlled programmed drug delivery over extended periods. It is used in place of eyedrops or ointments to lower elevated ocular pressure in glaucoma cases and is placed directly under the eyelid for 7 days. A low, constant therapeutic level of pilocarpine is maintained in the eye and reduced side effects (Figs. 3 and 4).

Alza has developed two therapeutic systems for oral and injectable use without the disadvantages of the corresponding conventional formulations. The OROS system is a new form of oral medication that contains a core of solid drug coated with an appropriate polymer membrane (Fig. 5) that is water-permeable

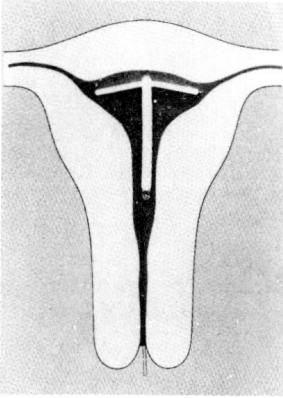

FIGURE 1. PROGESTASERT® intrauterine progesterone contraceptive system in the uterus.

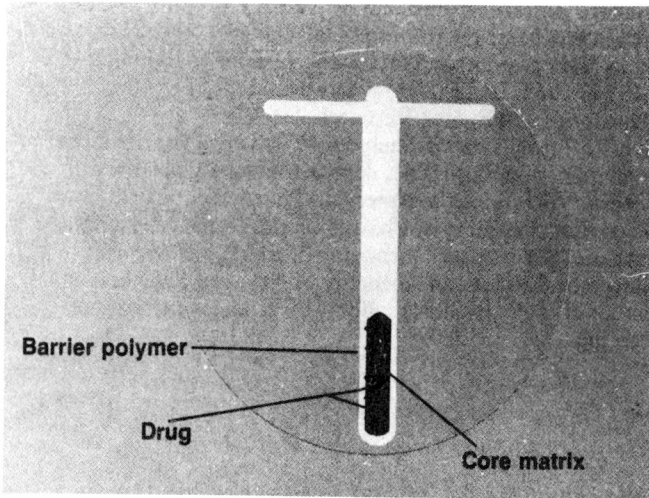

FIGURE 2. PROGESTASERT® system reservoir.

and has a small orifice. The membrane of this mini osmotic pump selectively admits water in which the drug dissolves. The internal pressure generated by entry of water forces the drug solution out of the orifice at a constant rate until all the drug has dissolved. The empty membrane capsule is excreted intact. Such systems can release drugs over a period of 24 h. The Chronomer system is a controlled-release norethisterone injectable that is active for 12 weeks.

One of the most sophisticated technologies for systemic treatment is the transdermal therapeutic system in which the released drug is absorbed through

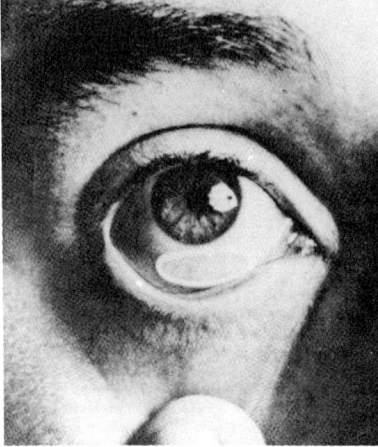

FIGURE 3. OCUSERT® ocular therapeutic system in place in lower cul-de-sac of eye.

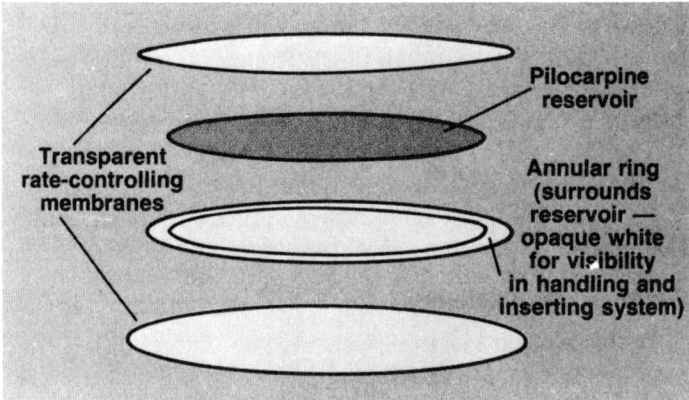

FIGURE 4. Exploded diagram of multilaminate structure of the OCUSERT® system.

intact skin into the circulatory system (Fig. 6). This is a major breakthrough in the history of medicine. One of the constraints in the application of this system is that the drug must be very potent and capable of being absorbed dermally. The first transdermal therapeutic system delivers scopolamine, a powerful anti-emetic drug. The system has multiple layers, providing a reservoir, controlled release, and adhesion to the skin. It can be placed behind the ear, where the skin is the thinnest and most permeable (Fig. 7). It is being evaluated for the control of nausea and vomiting induced by cancer chemotherapy and in the treatment of chronic dizziness. The system may prove valuable in the treatment of high blood pressure, angina, peripheral vascular disease, stroke, respiratory disease, mental deficiency in geriatric patients, and postmenopausal hormone deficiency.

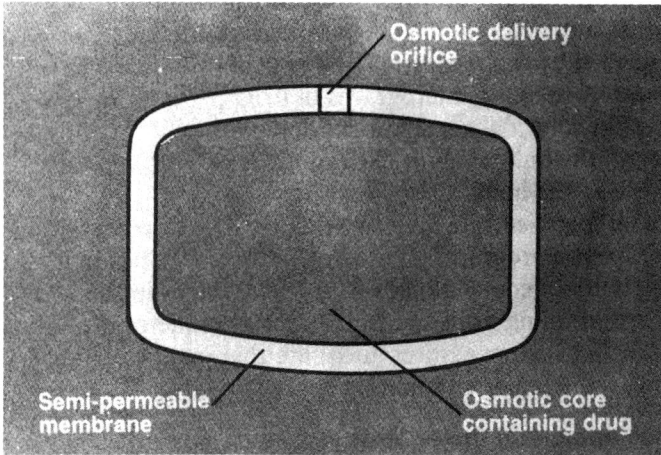

FIGURE 5.   Cross section of the OROS® system.

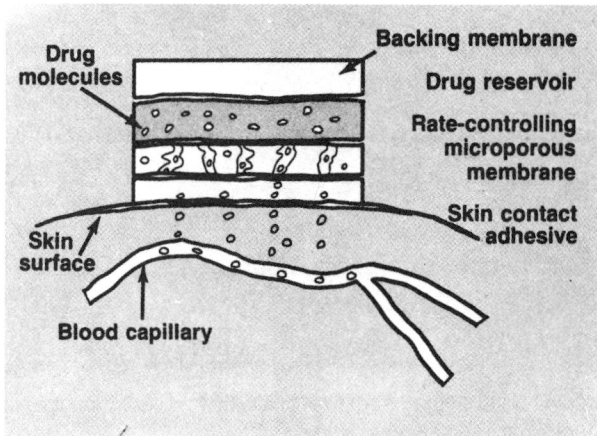

FIGURE 6.   Multilayer structure of transdermal therapeutic system.

Professor U. Zimmermann at the biochemistry department of Julich University, West Germany, has developed a drug delivery system in which the drug is incorporated into the patients' own red blood cells and the cells are then reintroduced by injection into the patient's body. The drug-containing red cells ("ghost cells") in circulation are indistinguishable from other red cells. They release the drugs when they decompose and die. Ghost cells are produced by applying a high voltage electric pulse through a solution containing the drug and the patient's red cells. The electric current "drills" a hole in the cell through which the drug enters. The cell quickly heals and the hole is sealed.

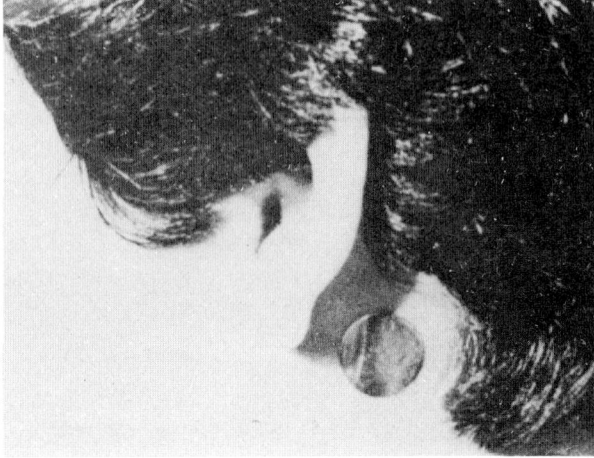

FIGURE 7. Transdermal therapeutic system (Scopolamine) in place behind the ear.

Methotrexate, an effective cancer drug, decomposes so rapidly that a high dose has to be given, which causes side effects, when administered conventionally. Using the ghost cells as carriers, methotrexate can be delivered directly to the liver without side effects. Ghost cells are now being modified so that they can be targeted to organs other than the liver and spleen. By adding into the ghost cell a protein molecule containing iron, it will be possible for doctors to hold these "magnetic ghost cells" where they want them to be and have them spill the drug at the target organs.

CHAPTER TWO

Design Parameters

R. L. COLLINS
Physics Department
The University of Texas at Austin, Austin, Texas

CONTENTS

1. INTRODUCTION

A properly designed controlled-release formulation will release the active agent according to the need, but equally important, it will *not* release excessive amounts under any foreseeable set of adverse conditions. The reservoir of active ingredient must be protected from deterioration, and the released substance must be at or translocate to the proper place to secure the action desired.

Considering the variability of conditions met in agronomic applications of pesticides, it is not surprising that the early successes of controlled-release technology have come in the pharmaceutical area. Controlled release of aspirin, antihistamines, and the like are relatively easy to attain. The temperature and other relevant conditions are predictable, and it only requires some experimentation among the formulations available to secure the desired result.

A second area that has achieved some success is the maintenance of a concentration of an agent at a surface. The insecticide in the Shell No-Pest® strip is embedded in a plastic strip, moves to the surface by diffusion, and then evaporates. Roach tapes have been designed using multilayer tape, with impervious lower layer, reservoir middle layer, and permeable upper layer. Flea and tick collars for pets dispense an embedded pesticide from the plastic collar and rely on diffusion, abrasion, and grooming to disperse the agent over the entire body. The rate of release is of course faster in the summer, when temperatures are high. In the hands of advertising people, however, this becomes a benefit since "in the summer, when insects are more of a danger, the flea collar releases extra amounts of active ingredient." Antifouling coatings based on rubber containing organotin complexes have had some success on naval vessels. On the other hand, the release of pesticides from rubber matrices for aquatic weed control has had little or no success.

Pheromones are employed to either attract males to traps or reduce mating through the confusion technique. The controlled release of tiny amounts of active agent is more difficult to achieve than in the previous examples, because of temperature variations and uncontrolled moisture (rain, perhaps dew). Most pheromones have unsaturated bonds and are highly susceptible to oxidative degradation. An antioxidant is therefore essential to protect the reservoir of active agent, and the antioxidant must vaporize less rapidly than the active agent.

General agronomic use of pesticides requires consideration of a much more complex set of environmental factors. Most natural enemies of crops spread their threat over many weeks or months. During this time, the soil and air temperatures are slowly warming, in addition to the daily temperature variation. Substances applied to foliage undergo more violent changes of temperature and also experience fully the effects of wind and weather.

It is evident that a great deal must be known about what controlled release needs to accomplish, if a suitable formulation is to be effected. It is, of course, *not necessary* to have a controlled-release formulation to find out the efficacy of such controlled release. All that is required is that the active agent be applied periodically, in the rates and over the times thought to be suitable. Surprises often occur. For example, the controlled release of 2,4-D from a metal carboxylic acid polymer formulation was found to be ineffective because of the soil absorption of small amounts of 2,4-D. This "soil burden" is a small thing when 2,4-D is applied conventionally but becomes burdensome when released very slowly. Another potential pitfall is that bacteria that feed on an active substance may proliferate around each source of that substance.

The testing of conventionally applied pesticides is a relatively simple manner. A row containing both target and nontarget species of plants is treated at different rates, typically using a "log rate," which means that the application rate falls off geometrically as one proceeds down the row. In equal increments of distance, the rate falls from 2 kg/ha to 1 to $\frac{1}{2}$ to $\frac{1}{4}$ to $\frac{1}{8}$, for example. The effective rate is then read by inspection. Compounds having a suitable combination of properties are readily noted from such trials. Some of the better candidates for controlled release may show poorly, on account of lack of persistence, but the test is simple, so it is much used. A single variable, the rate of application, is the only unknown quantity.

Contrast this with a controlled-release formulation. Several facets of the total package require optimization. The geometry, rate of release, mode of release, and effects of temperature, humidity and/or water, wind, soil type, and soil bacteria must be considered. The complexities of testing for optimization increase geometrically with the number of factors to be controlled. The best hope for holding the testing to a reasonable level of complexity lies in understanding what needs to be accomplished at various stages of the mission. Equally important, one must understand what must *not* happen. A sudden release of a larger-than-expected amount of a pesticide could have toxic effects on the species it is desired to protect. Hence, a close interaction between the formulations chemist, the agronomist, and the theoretical framework of controlled release are required to achieve success in this difficult field.

Persistence is a central factor in controlled release. Persistent pesticides, such as DDT, do not benefit from controlled release. From an agronomic viewpoint, except for rotation of crops, persistence is desirable. Environmentalists properly decry the translocation of persistent biologically active substances from the site of application, with the attendant risks of biological magnification through the food chain. However, the short-persistence pesticides favored by environmentalists tend to be toxic to handle, and expensive because multiple applications are needed. Controlled-release technology offers the promise of reducing toxicity and labor costs, but not the expense of the active chemical agent itself.

2. RELEASE MECHANISMS

A controlled-release formulation contains a reservoir of active agent and some means for releasing it to the environment. A zero-order rate means that the output rate is constant. An example is water evaporated from a cake tin, in which the rate of evaporation remains constant throughout. A first-order rate occurs when the rate declines with time, always being proportional to the amount of active agent still in the reservoir. A second-order rate means that the output rate is proportional to the square of the amount of agent remaining in the reservoir.

Diffusion-controlled release may be pseudo-first-order as in microcapsules or more complex when a substance in a polymeric matrix moves to the surface by diffusion and this diffusion is the rate-controlling step. Even if the motion to the surface occurs by diffusion, if the rate-controlling step is evaporation from the surface, the rate may be first-order.

Mathematically, if the total amount of active agent contained (and the amount that will eventually be released is M_∞, and M_t represents the amount released by time t, the release rate dM_t/dt is

Zero order: $\dfrac{dM_t}{dt}$ = constant

First order: $\dfrac{dM_t}{dt}$ = constant $\times (M_\infty - M_t)$

Second order: $\dfrac{dM_t}{dt}$ = constant $\times (M_\infty - M_t)^2$

Diffusion: $M_t \cong$ constant $\times M_\infty \times \sqrt{t}$ initially, more complex later
(if thin)

The rate law must be established experimentally. Further, the effects of temperature, moisture, and other relevant factors must be evaluated.

Higher temperatures usually speed the release. It is common that an Arrhenius law applies, such that the rate varies with temperature T (absolute temperature) as

$$\text{rate} = \text{constant} \times e^{-E/kT}$$

where E is an activation energy and k is the Boltzmann constant. Evaporative loss will also depend on wind velocity.

3. WHAT IS THE ACTIVE CONCENTRATION AT ANY TIME?

The actual concentration of active agent, released but not yet dispersed or degraded, is the balance between the input rate and the output rate. Whether or not this active agent is in the desired place depends on the design and application of the controlled-release formulation. For example, a granular formulation bearing an insecticide for soil insects and mixed into the soil at the appropriate depth should release the active agent where it will do the most good. A granular formulation designed to combat foliage-eating insects requires a sticking agent so that the active agent will be released onto the foliage.

The present concentration of active agent, $M(t)$, is the consequence of the previous input and loss rates. The calculations are simplified if a simple first-order rate is assumed for the loss rate, that is, if it is assumed that a half-life τ_0 exists such that a conventionally applied formulation falls to one-half its initial concentration in τ_0, to one-quarter its initial concentration in $2 \times \tau_0$, and to one-eighth its initial concentration in $3 \times \tau_0$. That is, if M_∞ is released instantly (no controlled release), the present concentration is

$$M(t) = M_\infty e^{-\lambda t}$$
$$= M_\infty e^{-(t \ln 2)/\tau_0}$$

The controlled-release formulation normally releases the active agent more slowly than the loss rate. A useful analogy is the leaky bucket. If the filling rate is g gallons per minute and the loss rate (through leaks) is $K \times G$ (where G is the total gallons in the bucket and K measures the size of the leaks), the bucket will fill until the rate of input is the same as the rate of loss, at which point $G = G^*$ (the equilibrium content):

$$KG^* = g$$

This volume of water in the bucket, $G^* = g/K$, will remain constant as long as g and K are unchanged.

The mathematical formulation for present concentration of an active agent, $M(t)$, when the loss is first-order with half-life τ_0 is

$$M(t) = \int_0^t \dot{M}_{t'} \, e^{-[(t-t')\ln 2]/\tau_0} dt'$$

where $\dot{M}_t'$ is the rate of release of the active agent. This rate of release depends on the methodology of controlled release and, in general, decreases with time. If the release rate law is known, exact solutions can sometimes be obtained. However, the effects of temperature and other environmental factors generally prevent obtaining exact solutions in closed form.

Even so, numerical integration can be used to obtain solutions. For example, suppose that $\dot{M}_t'$ has been determined experimentally for the relevant parameters of temperature, fraction of active agent released, and so on. By splitting time into, say, 1-h intervals and applying the expected time and temperature values (and other factors, as known) to the equations, a realistic solution for $M(t)$ can be had. Even small and inexpensive digital computers can be used to provide solutions by numerical integration.

For example, suppose that a controlled-release formulation releases its active agent at 25°C by first-order kinetics with half-life τ and that the active agent then is lost with half-life τ_0. At constant 25°C, the amount released in time t' is

$$M_{t'} = M_\infty e^{-\lambda t'} = M_\infty e^{-(t' \ln 2)/\tau}$$

where M_∞ is the total amount of active agent contained in the formulation prior to any release. The rate of release is

$$\dot{M}_{t'} = \frac{\ln 2}{\tau} M_\infty e^{-(t'\ln 2)/\tau}$$

and so the "present concentration" $M(t)$ is

$$M(t) = \int_0^t \frac{\ln 2}{\tau} M_\infty e^{-(t'\ln 2)/\tau} e^{-[(t-t')\ln 2]/\tau_0} dt'$$

$$= \frac{\tau_0}{\tau - \tau_0} M_\infty [e^{-(t\ln 2)/\tau} - e^{-(t\ln 2)/\tau_0}]$$

and it is readily seen that for $\tau \gg \tau_0$ the result is close to

$$M(t) = M_\infty \frac{\tau_0}{\tau} e^{-(t\ln 2)/\tau}$$

If no controlled release were used,

$$M(t) = M_\infty e^{-t\ln 2/\tau_0}$$

and so (1) the use of a first-order rate of controlled release simply changes the half-life from τ_0 to τ and (2) the initial amount applied, M_∞, must be increased to $M_\infty \tau/\tau_0$ to achieve the same initial present concentration.

Now consider temperature. Suppose that the active agent degrades at a rate

$$\lambda = \frac{\ln 2}{\tau_0} = A_0 e^{-E_0/kT}$$

where A_0 and E_0 are constants, experimentally measured. Suppose that the controlled-release formulation releases at a rate

$$\frac{\ln 2}{\tau} = A e^{-E/kT}$$

Hence, with changing absolute temperature T,

$$\tau_0 = \frac{\ln 2}{A_0} e^{E_0/kT}$$

$$\tau = \frac{\ln 2}{A} e^{E/kT}$$

Then, for a known time-temperature variation, $T = T(t)$, and the active "present concentration" is

$$M(t,T) = \int_0^t \dot{M}_{t'}(T) e^{-[(t-t')\ln 2]/\tau_0(T)} dt'$$

$$= M_\infty \ln 2 \int_0^t \frac{1}{\tau(T)} e^{-(t'\ln 2)/\tau(T)} e^{-[(t-t')\ln 2]/\tau_0(T)} dt'$$

where τ and τ_0 depend on T and hence on the temperature sequence $T(t)$. Hence, it would be necessary to first establish a temperature–time profile expected for the time span over which a controlled-release formulation is needed, and then numerically integrate over this time interval. Mathematical modeling can be very helpful, provided the parameters have been fully established. In this way, the effects of exceptionally hot, dry, cold, or rainy seasons can be modeled and the hazards and benefits noted.

The complications due to temperature are not all bad. The usual reduction of release rate with time can be partially compensated by the rise of average temperature as the growing season progresses. The toxic level of insecticides need not be maintained; simply exceeding the acute toxicity level once a day is usually adequate. In fact, maintaining the toxic level when all the insects are dead is quite wasteful.

4. SCALING LAWS

Everyone is familiar with the more rapid dissolution of fine powders compared with massive crystals. Similarly, large granules of a controlled-release formulation generally release their contents more slowly than small granules. The scaling law of release rate versus size depends on the mechanism. If evaporation from a surface is the rate-controlling step, then the rate is proportional to a^{-1}, where a is the diameter. If diffusion is rate-controlling, the rate is proportional to a^{-2}.

If the area changes as release occurs, the rate law becomes more complex. This can help. For example, internal surfaces molded into granules can keep the surface area from decreasing or even increase the surface area with time.

Scaling can help over only a limited range. If the size of the granule needed to give the desired rate were, for example, an inch in diameter, there would be a lot of space between granules and one would not obtain protection. Also, specialized machinery is needed when granule size falls outside the usual mesh sizes.

Geometry is a central factor in controlled release. For this reason, the use of liquid additives for controlled release does not seem to warrant consideration, since the geometry of a liquid is uncontrollable.

5. SUMMARY

It is probably worse than useless (and inevitably expensive) to attempt a controlled-release formulation without knowing in some detail *what* the formulation is supposed to do. Even if it is thought to be known, it is important that it be tried and verified. This can easily be done by repeated application according to the protocol.

The temperature and other environmental factors to which the formulation will be exposed in the course of its use, and possible extremes of these, affect the release rate of the formulation and the persistence of the active agent after release. Once these effects have been measured, mathematical modeling may help establish the needed rates of application and granule size. Also, potential hazards from unusual but possible weather conditions can be predicted.

The potential of controlled release of pesticides for agronomic applications is enormous. Nonpersistent pesticides may be made to masquerade as persistent pesticides at the site of application and so avoid the labor costs of frequent applications. The potential for problems is also large. The uncontrolled release of large concentrations of an active agent could be disastrous.

It seems clear that the combined efforts of agronomists and chemists will be required to achieve significant progress in this difficult area of technology.

CHAPTER THREE

Chemical Methods of Controlled Release

S. A. PATWARDHAN
National Chemical Laboratory, Pune, India

K. G. DAS
Regional Research Laboratory, Hyderabad, India

CONTENTS

1. INTRODUCTION

Controlled delivery of chemicals, though new to modern technology, occurs in nature. Nature operates through controlled-release systems to control contiguous competitive vegetation by maintaining optimum phytotoxic chemical concentrations. Oxygenation of blood and delivery and control of the passage of food and waste are other examples of controlled release. Development of controlled-release systems is an attempt to simulate nature's processes and to improve the efficiency of delivery and utilization of active ingredients at the target site. The drug industry is the first beneficiary of this technology. Major developments in the field of polymer science have paved the way to the development and commercial success of many controlled-release formulations for medical, pharamceutical, agricultural, forestry, public health, and veterinary applications.

A controlled-release system is prepared by either physically trapping or chemically binding the active ingredient to a suitable polymer, so that the optimum quantity of the active ingredient is released at a desired rate for a predetermined time at the site of action. Although the physical methods have considerable potential, they have certain drawbacks. To physically entrap an active ingredient by dissolution, dispersion, or microencapsulation, large amounts of inert polymer are required. In the chemical approach the pesticide is, directly or indirectly, covalently or ionically bonded to a preformed natural or synthetic polymer as a pendent group. Another approach is the preparation of a polymerizable pesticide monomer and its further polymerization alone or with a comonomer to a polymeric pesticide. Bifunctional pesticides are polymerized by intermolecular condensation to afford controlled-release polymers.

The chemical transformations help incorporate large amounts of active ingredients in comparatively small quantities of polymer matrix. The chemical bond between the pesticide and the polymer immobilizes the active ingredient until it is broken in the environment. The main limiting factor of this technique is that only pesticides and polymers having suitable mutually reactive functional groups are amenable to this approach. The active agent is released by the slow and sequential hydrolysis of the pesticide–polymer bond under the environmental conditions of application. This retro chemical synthesis is triggered by moisture, soil microbes, or natural light. It is necessary that the polymer–biocide bond must cleave more readily than any other bond in the biocide. Ester, anhydride, or amide linkages are the most useful. Both natural and synthetic polymers have found wide application in the design and synthesis of these systems. The great advantage of natural polymers over synthetic polymers lies in their biodegradability, low cost, nonpolluting nature, and availability from renewable resources.

The efficacy of these pesticide–polymer combinations depends on their rate of pesticide release. The release rates are controlled by the nature and stereochemistry of the pesticide–polymer linkage, the nature and presence of other functional groups, the level of substitution, hydrophobicity, and accessibility of the substrate–biocide bonds to the degrading agent. The extent of cross-linking and the degree of polymerization of the substrate also govern release rates.

Sustained-release drug dosage forms are prepared by converting the parent drug to prodrug derivatives (esters or ethers) or by synthesizing analogs. Derivatization is highly intuitive in nature, and the success depends on the expertise, experience, and orginality of the scientist. Each drug offers a challenge. Since definitive structure–reactivity correlations are not available, an overall rationale for chemical transformations of many classes of drugs has not evolved. Chemical modifications of a drug may alter its physicochemical properties to produce pharmacological and biochemical changes in the parent drug. Prodrug derivatives are believed to undergo enzymatic or nonenzymatic hydrolysis to the parent drug *in vivo*. Steroids, neuroleptics, β-lactams, antibiotics, antimicrobials, antimalarials, antihypertensives, and hypoglycemics have been chemically modified for sustained action.

2. TECHNIQUES

Fenone, tris(2,4-dichlorophenoxyethyl) phosphite, is the first example of an attempt to covalently link a pesticide to a polymer.[1] The potential use of polymers as a vehicle for controlled pesticide delivery was later demonstrated by Faerber, who prepared fungicidally active polymers by the homo- and copolymerization of *m*-chlorophenyl acetate.[2] In an early attempt to develop sustained-release formulations, alkyl resins were modified by the telomeric incorporation of 2,4-dichlorophenoxyacetic acid (2,4-D).[3] Most of these systems showed slow release rates, but the pattern of release was not desirable. The need to develop pesticides chemically attached to biodegradable polymers for long-term controlled release of pesticides was realized by Allan at the University of Washington, who has pioneered the work and developed the technique with considerable success. Harris at Wright State University has also made excellent contributions. Many controlled-release pesticide–polymer combinations have been prepared by the direct and indirect linking of preformed polymers to pesticides. Polymerization of monomeric pesticide derivatives has also been attempted.

2.1. Direct Bonding to Preformed Polymers

Pesticides have been directly bonded via covalent bonds as pendent substituents to preformed natural and synthetic polymers. For the preparation of such pesticide-polymer combinations, both the polymer and pesticide must contain accessible and mutually reactive functional groups. The new bond formed must degrade in the environment to release the active ingredient. The technique involves the reaction of a pesticide or its derivative with a polymer containing a suitable functional group. The reaction conditions needed to obtain a high degree of pesticide substitution must be carefully determined. Considerable work has been carried out by Allan and co-workers[4] on herbicides with carboxyl or hydroxyl groups and hydroxy-rich natural polymers. Some synthetic polymers containing hydroxyl and acid chloride groups have also been used for covalently binding pesticides. Ester, anhydride, and amide linkages are the most satisfactory covalent bonds for binding pesticides to polymers. Herbicides containing carboxyl groups can be converted to acid chlorides, which can react with polymers containing hydroxy or amino groups.

Cellulose present in many naturally occurring biodegradable polymers has been acylated with 2,4-dichlorophenoxyacetyl chloride (Scheme 1). The evaluation of the duration of herbicide release from these products in the soil was carried out from observations on the inhibition of the germination of lettuce seed that was sown daily.[4a] A product of Douglas fir bark and 2,4-D showed herbicidal activity for 100 days, while a kraft lignin-2,4-D combination was active for 170 days. Bark treated with 2,4-dichlorophenoxybutyric acid was

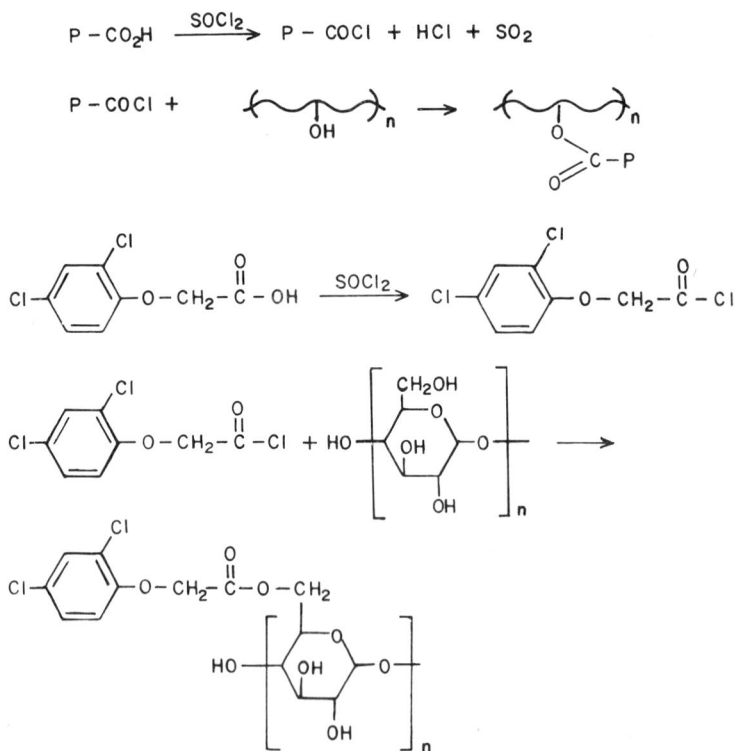

SCHEME I

reported to be effective for the long-term suppression of competitive vegetation in reforestation. Among naturally occurring polymers, cellulose, chitin, chitosan, alginic acid, lignin, starch, sawdust, and lignocellulosic bark have been found to be suitable for chemically bonding herbicides. Allan and co-workers have reviewed herbicide–natural polymer combinations.[5] Table 1 shows some of the natural and synthetic polymers that have been converted into herbicide–polymer combinations.

It has been shown[6] that heating a herbicide acid above its melting point in the presence of a natural polymer results in formation of a covalent bond be-

TABLE 1. Herbicide–Polymer Combinations

Herbicide	Polymer[a]	Liberated Herbicide %
Trichloroacetic acid (TCA)	Kraft lignin	6–9
	Bark	17.9
4-Chloro-2-methylphenoxyacetic	Kraft lignin	23–39
acid (2,4-D)	Bark	29–51
	Polyvinyl alcohol	56.1
2,4-Dichlorophenoxyacetic acid (2,4-D)	Cellulose	4–32
	Kraft lignin	15–39
	Bark	5–34
	Chitin	3–24
	Polyvinyl alcohol	42.8
	Polyethyleneimine	78.0
2,4,5-Trichlorophenoxy acetic	Cellulose	28–38
acid (2,4,5-T)	Kraft lignin	37–50
	Bark	17–24
2,2-Dichloropropionic acid (Dalapon)	Kraft lignin	10.6
	Bark	20–33
	Polyvinyl alcohol	45.0
2-(2,4,5-Trichlorophenoxy propionic	Kraft lignin	12–43
acid (Silvex)	Bark	8–33
4-(4-Chloro-2-methylphenoxy)	Kraft lignin	12–42
butyric acid (MCPB)	Bark	8–44
4-(2,4-Dichlorophenoxy) butyric	Kraft lignin	29–40
acid (2,4-DB)	Bark	20–50
4-(2,4,5-Trichlorophenoxy) butyric	Kraft lignin	22
acid (2,4,5-TB)	Bark	11.5
4-Amino-3,5,6-trichloropicolinic	Kraft lignin	7–6
acid (Picloram)	Bark	11.0
	Sawdust	8.6

tween them. 2,4-Dichlorophenoxyacetic acid and Douglas fir bark heated together at 150° for 3 h yielded a product containing 20% herbicide. These polymer
esters have also been synthesized by exchange reactions. Thus heating 2,4-D
with cellulose acetate at 150° for 3 h gave a product containing 33% covalently
bonded 2,4-D. The anhydride forms of herbicidal acids have also been used to
esterify polymers.[7] A synthetic polymer such as polyvinyl alcohol has been
acylated with 2-methyl-4-chlorophenoxyacetyl chloride. The product inhibited
lettuce seed germination for 100 days. Acid chlorides of 2,4-D, 2,2-dichloro-
propionic acid, and 4-chloro-2-methyl-phenoxyacetic acid have been treated
with many synthetic polymers such as polyvinyl alcohol and polyethyleneimine
to give products containing up to 50% hydrolyzable active ingredient.

Jakubka and Busch[8] prepared synthetic bioactive polymers by the metathetical
reaction of triethylammonium salts of 2,4-D, 2(4-chloror-2-methylphenoxy)pro-
pionic, 4-(4-chloro-2-methylphenoxy)butyric, 4-(2-4-dichlorophenoxy)butyric,
2-(2,4-dichlorophenoxy)propionic, 8-quinolinoxyacetic, and 4-chloro-2-methyl-
phenoxyacetic acids, with the chloromethylated derivative of divinyl benzene-
styrene copolymer.

Mixed anhydrides of pesticides and polymers can be easily synthesized. 2,4-D
forms anhydrides with polyacrylic and polysebacic acids.[9] These products did
not show significant herbicidal efficacy. However, the duration of activity increases by vicinal enclosure of the anhydride link by a hydrophobic aromatic
nucleus. These products are best suited for short-term control (Scheme 2).

Pesticides with hydroxyl groups react with synthetic polymers containing
pendent acid chlorides. Several pesticides have been treated with polyacryloyl
chloride and polymethacryloyl chloride (Scheme 3). Pentachlorophenol reacts
with polyacryloyl chloride. A series of antifouling polymers were prepared by
treating phenarsazine, 2,4,6-trichlorophenol, tri-n-butyltin hydroxide, 1-hydroxy-
imidazoles, and 1-hydroxyimidazole-3-oxides with polyacryloyl chloride or
polymethacryloyl chloride[10,11]. Styrene–maleic anhydride copolymer reacts
with pentachlorophenol, phenazo-2-naphthol, p-hydroxy-phenylazo-2-naphthol,
phenarsazine chloride, and tributyltin hydroxide[11]. Some typical pesticides and
polymers suitable for chemical binding are shown in Table 2.

Rzaev[12] has reported the synthesis of many antifouling copolymers by treating polymers containing anhydride groups with trialkyl stannanoles (R_3SnOH)
and hexaalkyl distannoxanes ($R_3SnOSnR_3$). Polymaleic anhydride, copolymers
of maleic anhydride with vinyl monomers (styrene, vinyl acetate, vinyl chloride,
vinyl triethoxy silane), were used as anhydride polymers. The organotin groups
were found to be uniformly distributed in the backbone of the copolymers.
Uniformly alternating copolymers of bistrialkyl stannyl maleate have also been
prepared. Polyfunctional macromolecules prepared from hexaalkyl distannoxanes
and terpolymers containing maleic anhydride were observed to be capable of
cross-linking (Scheme 4).

SCHEME 2

SCHEME 3

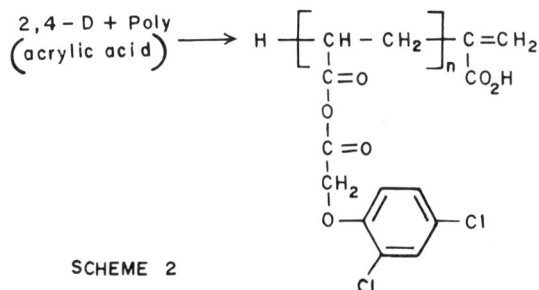

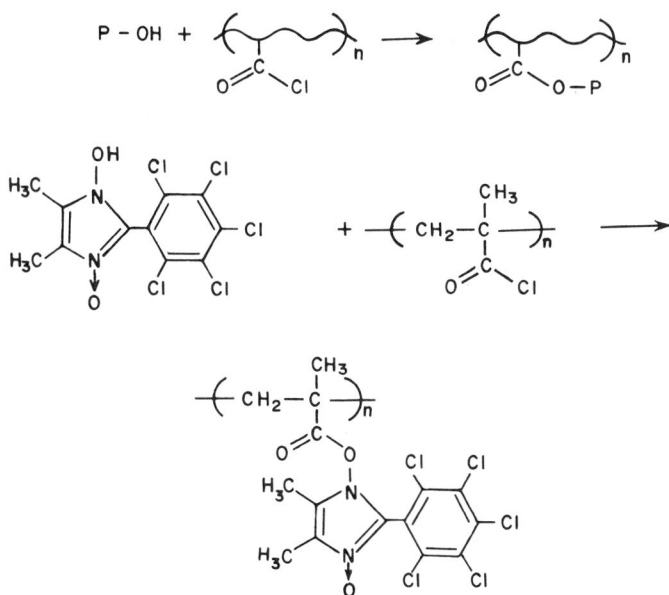

Pesticidal aldehydes have been treated with polymers containing hydroxy groups. An acetal linkage is formed to bind the pesticide and polymers. 2,6-Dichlorobenzaldehyde-generating polymers were synthesized by Schacht and co-workers[13] by making use of the above reaction (Scheme 5). A polyvinyl acetal copolymer was prepared by treating 2,6-dichlorobenzaldehyde with polyvinyl alcohol in the presence of acid. The degree of acylation was as high as 80%. The product was found to be stable in water at 25° (pH 1).

Pesticides can be covalently linked by amide linkages. Bioactive synthetic polymer with amide linkages have been prepared from phenoxy herbicide acids

TABLE 2. Direct Bonding

Pesticide	Polymer[a]
Pentachlorophenol	Polyacryloyl chloride
Phenylazo-2-naphthol	Polymethacryloyl chloride
p-Hydroxyphenylazo-2-naphthol 2,4,6-Trichlorophenol	Polymaleic anhydride
Triethyltin hydroxide	Copolymer of maleic anhydride with vinyl monomer
Tri-n-butyltin hydroxide	
1-Hydroxyimidazole	
1-Hydroxyimidazole-3-oxide	

[a]Containing acid chloride or anhydride group.

and aminopolysaccharides such as chitosan or fish wastes.[6] These natural polymers contain reactive amino carboxyl groups. 3,6-Dichloro-o-anisic acid has been covalently bonded to fish waste. A product containing 35% active ingredient has been prepared. Comparative studies on the efficacy of the controlled-release herbicide combination and the free herbicide indicated that the former suppressed the growth of vegetation in pots five times longer than the latter when equivalent amounts were applied. Crab, shrimp shell, and fungal hyphae have also been used as natural polymers.[14] When 2,4-D is heated with chitosan in the ratio $2:1$ at $195°$ for 1h, a heat-induced covalent bond formation afforded a product containing 45% herbicide.

Low- and medium-molecular-weight polyethyleneimine have been acylated with 2,4-D to give a series of polymers with different degrees of substitution. The release rate in the soil showed wide variation with no correlation with molecular weight. However, the degree of substitution of herbicide had some control on the rates of release. Faster release rates were observed with less sub-

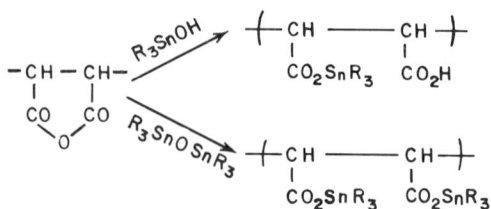

SCHEME 4

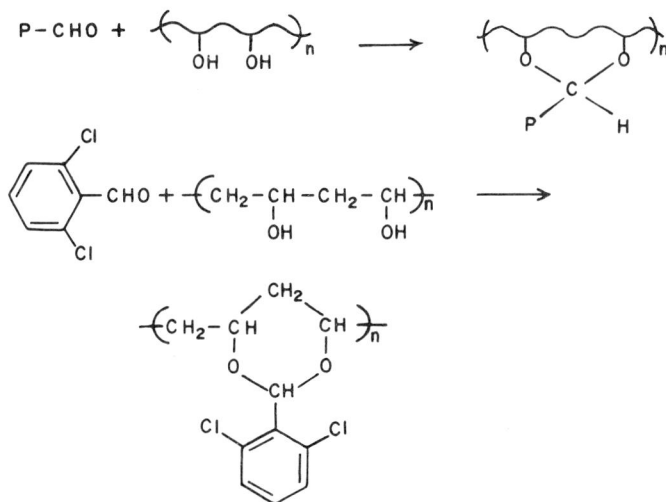

SCHEME 5

stitution. It was reported that amide bonds are cleaved more readily than ester bonds. The hydrophilicity of the product affects the accessibility of the hydrolytic enzyme to the binding linkage. This is more important than the bonding energy in determining release rates. Polyethyleneimine-5-alkyl-dithiocarbamates having controlled-release properties have been reported[15,16]. 2,4-D and amino-methylated resin from divinylbenzene copolymer react to give a product with pendent amide linkage. The reaction takes place in the presence of dicyclohexyl-carbodiimide. Amino-methylated derivatives of styrenic copolymer can be acylated with herbicide acids.

Herbicides containing amino groups have been covalently attached to natural and synthetic polymers. A controlled-release herbicide was prepared by heating in the dry state lignin sulfonic acid that had been neutralized with 3-amino-1,2,4-triazole. The product showed herbicidal activity, indicating that it contains hydrolyzable sulfonamide or imide linkages.

2.2. Indirect Bonding to Preformed Polymers

If the active agent and polymer contain functional groups that are not mutually reactive for direct covalent bonding, bridging between the groups can be effected by interposing a multifunctional entity. The bond between the active agent and the binding compound must be more easily cleaved in the environment than the link between the polymer and the binding molecule. Pesticides containing hydroxy or amino groups have been successfully attached by covalent bonds with the aid of bridging molecules to polymers containing hydroxy or amino groups. Initially equimolar amounts of the pesticide and the bridging compound are made to react. The product formed is treated with the polymer. Side reactions and cross-linking of the polymer must be avoided. Polyfunctional molecules such as toluene diisocyanate, carbonyl chloride, phosphoryl chloride, and silicon chloride have been successfully used for bridging.

Pentachlorophenol was covalently attached to softwood kraft pulp that had been treated previously with cyanuric chloride by s-triazine bridges. Allan and Halabisky[17] have synthesized several systems from 2,4-D, pentachlorophenol, or 4-amino-3,5,6-trichloropicolinic acid and proteins or polyethyleneimine using acrolein and cyanuric chloride. Metribuzin adducts with diisocyanates react with polyvinyl alcohol to give linear and cross-linked copolymers with different amounts of metribuzin substitution. Rate-of-release studies in water indicated that the herbicide release is faster in linear than in cross-linked polymers. The soil mobility of the released active ingredient was found to be considerably reduced, whereas the residual phytotoxicity was appreciably more.[18]

Metribuzin, 4-amino-6-(1,1-dimethylethyl)-3-(methylthio)-(1,2,4)-triazin-5(4H)-one, was bridged to polyvinyl alcohol[19] with toluene diisocyanate. The use of bridging molecules to prepare controlled-release formulations of the plant hormones belonging to the cytokinin group has been reported.[20] The polymer-bridging compound bond was formed before attaching the active agent. The second bond is cleaved during the hydrolytic release of the plant hormone. Phosgene reacts with a polyhydroxypolymer to give a chlorocarbonate. The seconddary amino group of cytokinin can react with the chlorocarbonate to produce a decomposable urethane bridge (Scheme 6). Hydrolytic cleavage results in the elimination of carbon dioxide and the release of cytokinin. More examples of pesticides, polymers, and bridging molecules are indicated in Table 3.

2.3. Polymerization of Pesticide Monomers

The work on the synthesis of polymeric pesticides from pesticide monomer derivatives was initiated independently by Faerber[2] and Kostanyan.[21] The main requirement in this approach is that the pesticide molecule must contain a suitable functional group which can be converted to a polymerizable derivative.[22] This limits the number of pesticides that can be converted to polymerizable monomers. Pesticides containing carboxyl, hydroxyl, sulfhydryl, and amino groups can be converted to polymerizable vinyl derivatives. Pesticides that do not possess any one of these functional groups are not amenable to this approach.

Most of the work in this area has been on herbicides. The herbicide is first converted to a monomer by introducing a vinyl group in the molecule. The herbicide monomer is then polymerized. Considerable work on this approach has been carried out by Harris and co-workers at Wright State University. These

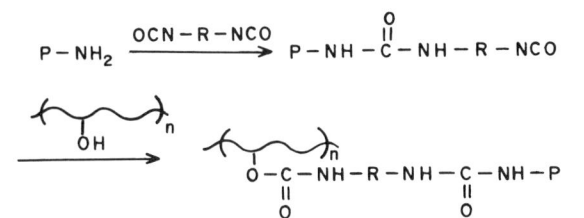

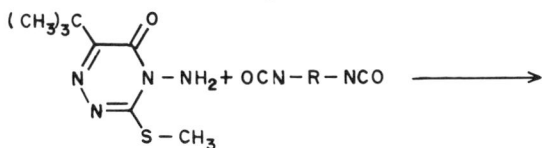

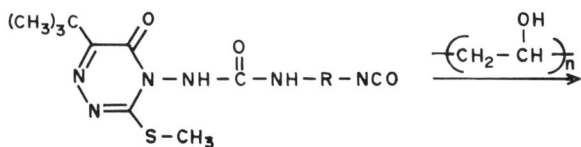

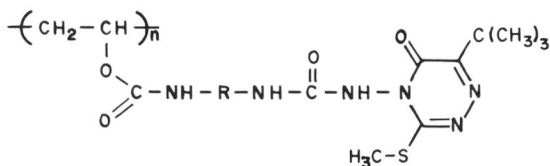

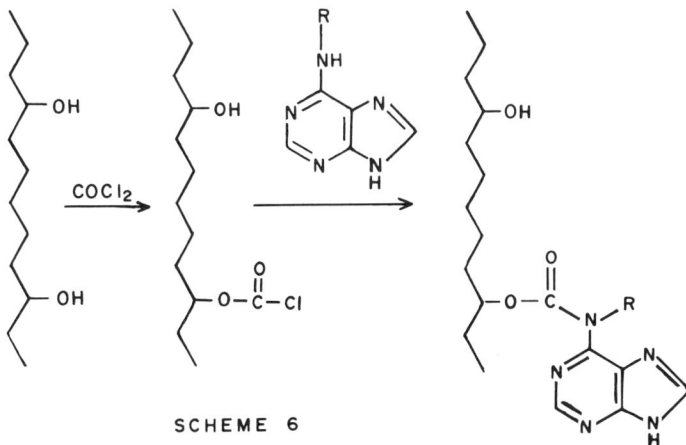

SCHEME 6

R= CH₂·C₆H₅

TABLE 3. Indirect Bonding via Bridging Molecule

Pesticide	Polymer	Bridging Molecule
Pentachlorophenol	Softwood kraft lignin	Cyanuric chloride
2,4-D	Proteins	Phosphoryl chloride
4-Amino-3,5,6-trichloropicolinic acid	Polyethyleneimines	Silicon chloride
Metribuzin	Polyvinyl alcohol	Toluene diisocyanate
Cytokinins		Carbonyl chloride

monomers usually contain easily hydrolyzable bonds such as ester, amide, urea, or carbamate. From the polymerized macromolecule, the herbicide is released mainly by hydrolysis.

2.3.1. Preparation of Pesticide Monomers

Herbicides containing carboxylic groups are converted to vinyl derivatives by vinyl exchange reaction between the carboxyl group of the herbicide and vinyl acetate in the presence of mercuric acetate and sulfuric acid catalyst[23]. The vinyl group of the acetate is transferred to the pesticide, thereby replacing the acidic hydrogen (Scheme 7).

Vinyl 2,4-dichlorophenoxyacetate, vinyl (4-chloro-2-methylphenoxy)acetate, vinyl (2,4,5-trichlorophenoxy)acetate, and vinyl 4-(2,4,5-trichlorophenoxy)-butyrate have been prepared and polymerized to give polymers with herbicide as a pendent substituent directly attached to the polymeric backbone. Their rate of hydrolysis is extremely slow. It has been demonstrated that the rate of hydrolysis can be enhanced by increasing the distance between the polymeric main chain and the ester group. The hydrolysis is slow, owing to the hydrophobicity of the polymer. The rate of release of 2,4-D will be related to the rate of cleavage of the covalent bond between the pendent pesticide substituent and the polymeric main chain. This again depends on the strength and chemical nature of the covalent bond and on the chemical and stereochemical properties of the vicinal polymer backbone. Increasing the hydrophilicity of the backbone should enhance the hydrolytic cleavage. This has been demonstrated by Neogi[24] and Wilkins.[25] Hydrophilicity can be introduced by copolymerization of the herbicide with an appropriate amount of acrylic acid.

Many polymerizable herbicide derivatives can be prepared by treating the herbicide acid with alcohols containing a vinyl group.[26] Reagents such as dicyclohexylcarbodiimide catalyze the reaction. In a two-step process, the acid is first converted to the acid chloride by thionyl chloride. The highly reactive acid

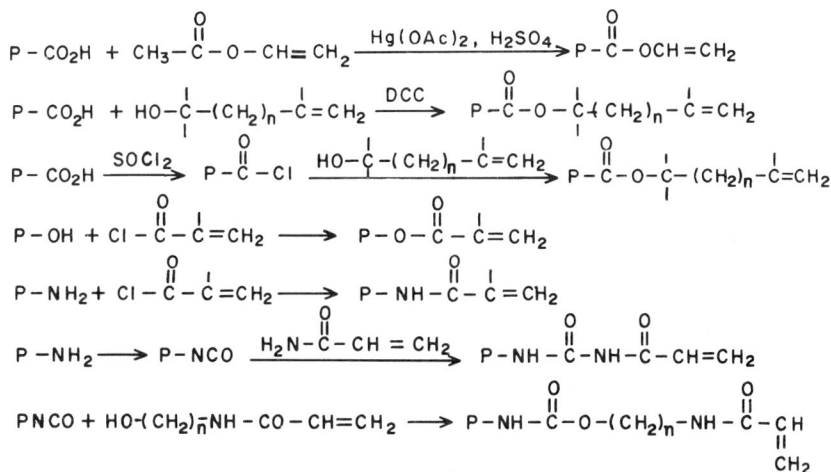

$$P-CO_2H + CH_3-\overset{O}{\overset{\|}{C}}-O-CH=CH_2 \xrightarrow{Hg(OAc)_2,\ H_2SO_4} P-\overset{O}{\overset{\|}{C}}-OCH=CH_2$$

$$P-CO_2H + HO-\overset{|}{\underset{|}{C}}-(CH_2)_n-\overset{|}{C}=CH_2 \xrightarrow{DCC} P-\overset{O}{\overset{\|}{C}}-O-\overset{|}{C}(CH_2)_n-\overset{|}{C}=CH_2$$

$$P-CO_2H \xrightarrow{SOCl_2} P-\overset{O}{\overset{\|}{C}}-Cl \xrightarrow{HO-\overset{|}{C}-(CH_2)_n-\overset{|}{C}=CH_2} P-\overset{O}{\overset{\|}{C}}-O-\overset{|}{C}-(CH_2)_n-\overset{|}{C}=CH_2$$

$$P-OH + Cl-\overset{O}{\overset{\|}{C}}-\overset{|}{C}=CH_2 \longrightarrow P-O-\overset{O}{\overset{\|}{C}}-\overset{|}{C}=CH_2$$

$$P-NH_2 + Cl-\overset{O}{\overset{\|}{C}}-\overset{|}{C}=CH_2 \longrightarrow P-NH-\overset{O}{\overset{\|}{C}}-\overset{|}{C}=CH_2$$

$$P-NH_2 \longrightarrow P-NCO \xrightarrow{H_2N-\overset{O}{\overset{\|}{C}}-CH=CH_2} P-NH-\overset{O}{\overset{\|}{C}}-NH-\overset{O}{\overset{\|}{C}}-CH=CH_2$$

$$PNCO + HO-(CH_2)_n-NH-CO-CH=CH_2 \longrightarrow P-NH-\overset{O}{\overset{\|}{C}}-O-(CH_2)_n-NH-\overset{O}{\overset{\|}{C}}-\underset{\underset{CH_2}{\|}}{C}H$$

SCHEME 7

chloride reacts in the presence of pyridine with an alcohol containing a vinyl group. Acid chlorides also react with the preformed sodium salts of alcohols.

Pesticides containing hydroxyl groups have been converted to their vinyl derivatives. The hydroxyl group of the pesticide is treated with an acid or acid chloride containing a vinyl group. Thus pentachlorophenol reacts with methacryloyl chloride to give pentachlorophenyl methacrylate, which has been polymerized to an effective antifouling polymer. Similarly, 2,3,5-trichloropyrid-4-yl methacrylate has been prepared from 2,3,5-trichloro-4-hydroxypyridine and methacryloyl chloride.

Metribuzin has been converted to a polymerizable form with a hydrolyzable amide bond. The amino group has been treated with acryloyl chloride or methacryloyl chloride to produce N-acryloyl metribuzin and N-methacryloyl metribuzin, respectively. These monomers have been further polymerized.[23] Metribuzin isocyanate has been converted to vinyl monomer by treating with N-(2-hydroxyethyl)-acrylamide.[27]

Some of the other pesticide monomers reported in the literature are glycidyl (4-chloro-2-methylphenoxy)acetate, N-(4-chloro-2-methylphenoxy)acetyl aziridine, pentachlorophenyl methacrylate, 2,3,5-trichloropyrid-4-yl methacrylate, and N-hydroxymethyl (2,4-dichlorophenoxy)acetamide.

2.3.2. Polymerization of Monomers

Pesticide monomers have been polymerized by solution, bulk, and emulsion free-radical techniques.[22] The simplest method is bulk polymerization. The monomer is heated with about 1% of a free radical initiator (azobisisobutyronitrile) for several hours. The polymerization is rarely complete, and hence the polymer invariably contains unreacted monomer, which is difficult to remove. Since the polymerization is exothermic, there are problems of heat control and the danger of explosions. Vinyl (2,4-dichlorophenoxy)acetate is mixed with 1% azobisisobutyronitrile and slowly heated to 70°. The heating is continued for 3 h, and the product is dissolved in chloroform. Precipitation with hexane followed by extraction with ethanol give the polymer.

Solution polymerization is carried out in suitable solvents. The solvent should not interfere with the polymerization and should take up the heat of polymerization. Low-molecular-weight products are generally obtained. It is difficult to completely remove the solvent from the polymer.

In emulsion polymerization the monomer is dispersed in water by an emulsifier, and the dispersed monomers migrate into soap micelles. Redox-type initiator radicals are generated in the water diffuse into the micelles, which are swollen with monomer molecules. High-molecular-weight polymers are formed that are suitable for direct use in formulations. A few examples of pesticide polymers prepared from vinyl pesticide monomers are presented in Table 4.

Both homopolymerization and copolymerization approaches have been adopted for polymerization of herbicide monomers. In copolymerization the pesticide monomer is polymerized together with another molecule with a different functional group that will enhance the rate of hydrolysis of the bond that links the pendent pesticide moiety to the backbone of the polymer chain. Monomers containing herbicides, fungicides, and antifoulants have been prepared. It has been shown that the presence of hydrophilic groups enhances the release rates and thereby increases the biological control. The pendent carboxyl group in acrylic acid-2,3,5-trichloro-4-pyridyl methacrylate copolymer catalyzes the hydrolysis of the vicinal ester bond. It was demonstrated that the rate of hydrolysis of the herbicide–polymer bond is enhanced by increasing the length of the pendent side chain, since the ester would be away from the hydrophobic backbone and sterically less hindered. Many homopolymers of herbicides do not undergo hydrolysis. Copolymers containing varying amounts of hydrophilic monomers such as methacrylic acid and trimethylamine methacrylimide[28] are hydrolyzable.

Harris et al.[29] have developed many controlled-release aquatic herbicide systems. The 2-acryloyloxyethyl and 2-methacryloyloxyethyl esters of 2,4-D have been copolymerized with methacrylic acid and 2-hydroxyethyl methacrylate to give the corresponding copolymers. These copolymers have been found to control many aquatic weeds. A series of copolymers of 2-methacryloyloxyethyl 2,4-dichlorophenoxyacetate containing varying amounts of glyceryl methacrylate were prepared in 94–96% yields.[30] Spectrophotometric determination of the amount of 2,4-D released showed that the initial rate of hydrolysis depends on the amount of hydrophilic comonomer present. The copolymer appears to become more hydrophilic as the herbicide is released from the polymer.

TABLE 4. Pesticide Polymers from Monomers

REACTANTS	PESTICIDE MONOMER AND COMONOMER	POLYMER
2,4-D + **VINYL ACETATE**	2,4-D vinyl acetate monomer structure	2,4-D vinyl acetate polymer structure
2,4-D +4-ACRYLOYLOXY BUTYL ALCOHOL	monomer structure	polymer structure
PENTACHLOROPHENOL + **METHACRYLOYL CHLORIDE**	monomer structure	polymer structure
METRIBUZIN + **ACRYLOYL CHLORIDE**	monomer structure	polymer structure
METRIBUZIN ISOCYANATE + **ACRYLAMINE**	monomer structure	polymer structure

37

TABLE 4 (continued)

METRIBUZIN ISOCYANATE +
2-AMINOETHYL ACRYLATE

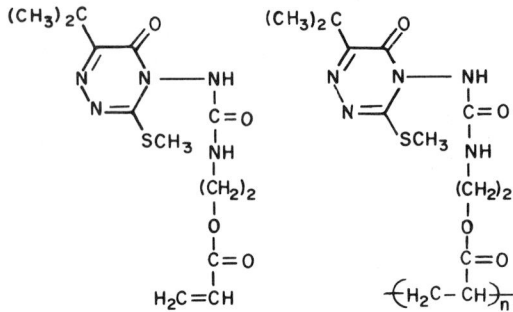

2-4-D + 2-HYDROXY ETHYL
ACRYLATE

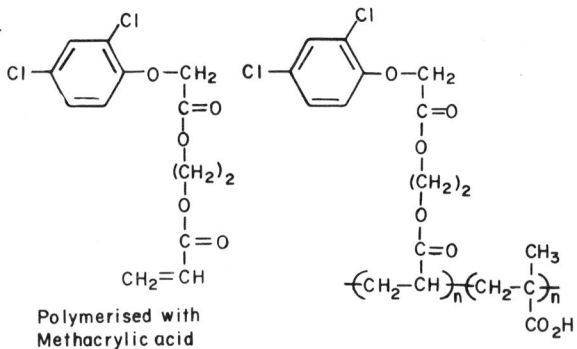

Polymerised with
Methacrylic acid

SILVEX + 2-HYDROXY ETHYL
ACRYLATE

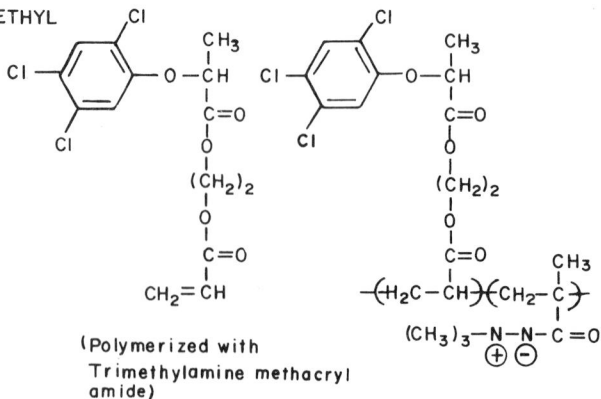

(Polymerized with
Trimethylamine methacryl
amide)

Increasing the ratio of hydrophilic comonomer in the system leads to increased herbicide release rates. The rate of hydrolysis also increases with time. This has been attributed to intramolecular interactions of the unhydrolyzed ester groups with the free carboxyl groups formed on the chain by hydrolysis. Pentachlorophenyl methacrylate has been prepared from pentachlorophenol and methacryloyl chloride. The product obtained after polymerization showed long-term antifouling action. The growing interest in acrylic polymers containing pendent hydrolyzable toxic groups as fungicides, pesticides, antifouling coatings, and wood preservatives has led to the synthesis of homopolymers of tri-n-butyltin methacrylate and tri-n-butyltin acrylate, their copolymers, and terpolymers with other monomers.[31,32] Some of the other organometallic groups linked to the polymer backbones are tributyltin, tripropyltin, trimethyltin, triphenyltin, and tribenzyltin. The propyl and butyl derivatives were reported to be most effective as antifouling agents. The activity of the polymer was observed to be enhanced by the introduction of two or more different organometallic groups in the backbone. Poly(tri-n-butyltin methacrylate-Co-tri-n-propyltin methacrylate-Co-methyl methacrylate) showed good antifouling action for several years. Antifouling paints contain poly(tri-n-butyltin methacrylate-Co-methyl methacrylate) and poly(tri-n-butyltin methacrylate-Co-tri-n-propyltin methacrylate-Co-methyl methacrylate).[33] The addition of cross-linking agents results in the conversion of both the polymers to hard paint films. These resins combine the transparency and film-forming properties of acrylics with the antifouling activity of organometallic compounds. When these organometallic acrylates were copolymerized with fluoroalkyl acrylates, water- and oil-repelling properties were imparted to their coatings. Hydrophilic comonomers such as 2-hydroxyethyl acrylate have also been copolymerized with organotin acrylate monomers.[34] The hydrophilic group produces drag-reducing coatings with antifouling properties.

2.4. Newer Methods

It is obvious that in all the pesticide–polymer combinations described so far in this chapter, the backbone of the polymer is inactive and wasteful. In a novel approach, Allan and Neogi[35] converted bifunctional pesticides to polymers that contain only the bioactive material. The active ingredient is released from such systems by degradation of the main polymer chain rather than by cleavage of pendent substituents. The release of the active ingredient has to take place by endwise degradation. Hence, increasing the degree of polymerization of the linear polymer should lead to reduction in the rate of release. The release rates therefore depend on the degree of branching. One of the requirements is that the bonds formed during polymerization must be hydrolyzable. The aromatic polyamides have been reported to be very stable and hence are not desirable. An exception to this is poly(3,5,6-trichloropicolinamide), which is an active herbicide. It has been reported that a heteroatom such as nitrogen of a heterocyclic aromatic ring anchimerically assists the hydrolysis.

Only a few pesticides contain more than one functional group that could be polymerized by intermolecular condensation to give satisfactory degradable polymeric pesticides. Polymers in which pesticides are incorporated in the main chain are synthesized by the homopolymerization of pesticides containing two different groups or by the copolymerization of pesticides having two identical groups with appropriate bifunctional comonomers (Scheme 8). Thus pesticides containing carboxyl and amino groups have been homopolymerized to give polyamides. A pesticide containing two carboxyl groups can be copolymerized with a dialcohol to afford a polyester.

Akagane and Allan [36] have reported other examples of bioactive polymers of a similar type with antifouling properties. Inorganic polyanhydrides

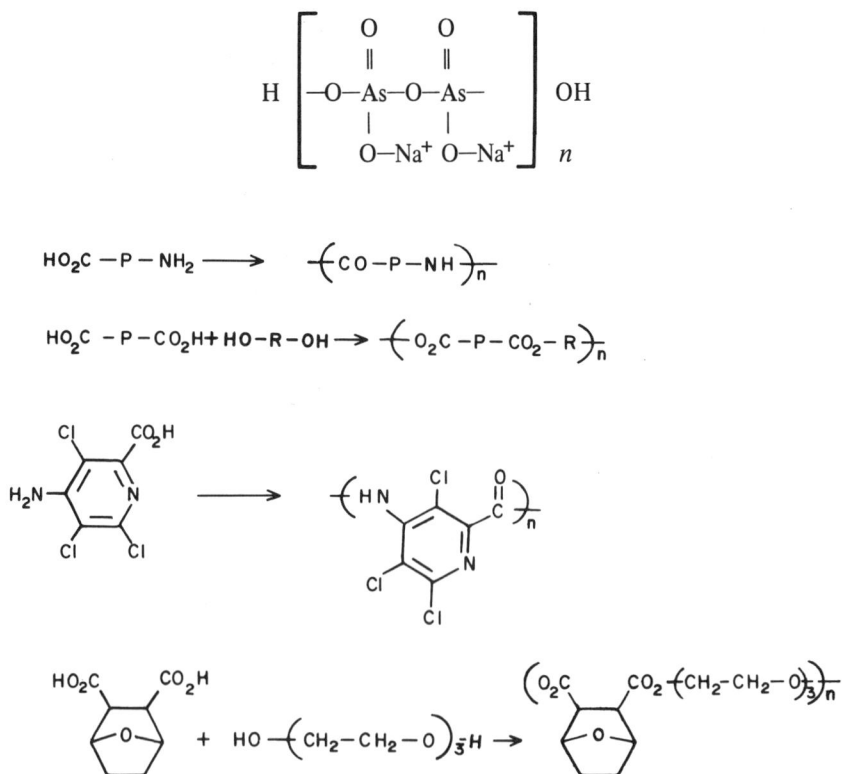

SCHEME   8

were prepared by the thermal dehydration of monosodium arsenate, which was first heated to 250° for 1 h and then at 700° for 6 h alone or with elemental sulfur. A sulfur monomer content of 5-10% was found to be necessary to stabilize the antifouling properties for a period of 30 days. Satisfactory antifouling activity has been reported for the homopolymer polyphenarsazine and the copolymer of 1-hydroxy-2-(4-hydroxyphenyl)-3-oxido-4,5-dimethylimidazole and terephthalic acid. The degree of polymerization is kept low so that sufficient polymer chain ends are available for triggering the endwise hydrolytic degradation.

Attempts have been made to design controlled-release systems in which more than one pesticide has been chemically bonded to one polymer. For weed control there is a need to apply more than one chemical. Synergistic effects are also known. Instead of applying two distinct controlled-release pesticide–polymer combinations, it will be advantageous if one controlled-release system can deliver both the required active agents from the same product. Akagane[37] has developed a low-cost lignin-based complex herbicide system. The phenolic hydroxyl groups in kraft lignin are etherified by cyanuric chloride to afford dichloro-s-triazinyl derivative (I) (Scheme 9). 2,4-D is treated with diethylenetriamine (II). The active halogen in I is then replaced by the free amino group of II to give III. The free amino groups in the resulting lignin complex are further made to react with 3,4,5,6-tetrachlorophthalic anhydride and tributyltin hydroxide to yield the complex multiherbicide polymer combination IV.

Antifouling products with improved characteristics were obtained by the action of polyacryloyl chloride with two or more active agents selected from phenarsazine chloride, 1-hydroxy-imidazoles and their 3-oxido analogs.[38] The mutual interactions of these active agents on a single polymer gave better antifouling action in the marine environment than mixtures of the corresponding polymers containing a single toxicant.

In an *ab initio* synthesis of pesticide–polymer combinations it has been conceived that it should be possible to synthesize a polymer containing pendent groups that would slowly hydrolyze to release the active pesticide component.[38] In this approach, a controlled-release biocide–polymer combination is synthesized without using the biologically active biocide. This technique is advantageous when the pesticide is difficult to synthesize, unstable, or hazardous to handle due to its high mammalian toxicity. Octamethyl pyrophosphate has a high mammalian toxicity and is used as a rodenticide. Its delayed action has been explained as due to oxidation to an amine oxide or a hydroxymethyl derivative, which takes place during its latent period. The resulting electronic change facilitates the hydrolysis of pyrophosphate bonds and the reaction with chymotrypsin. There is ample evidence to support this hypothesis. Assuming that the hydrolytic products are physiologically active, it is interesting to synthesize a polymer containing this pendent group which can hydrolyze to release the active pesticide component. With this objective a toxic group such as phosphoramide has been

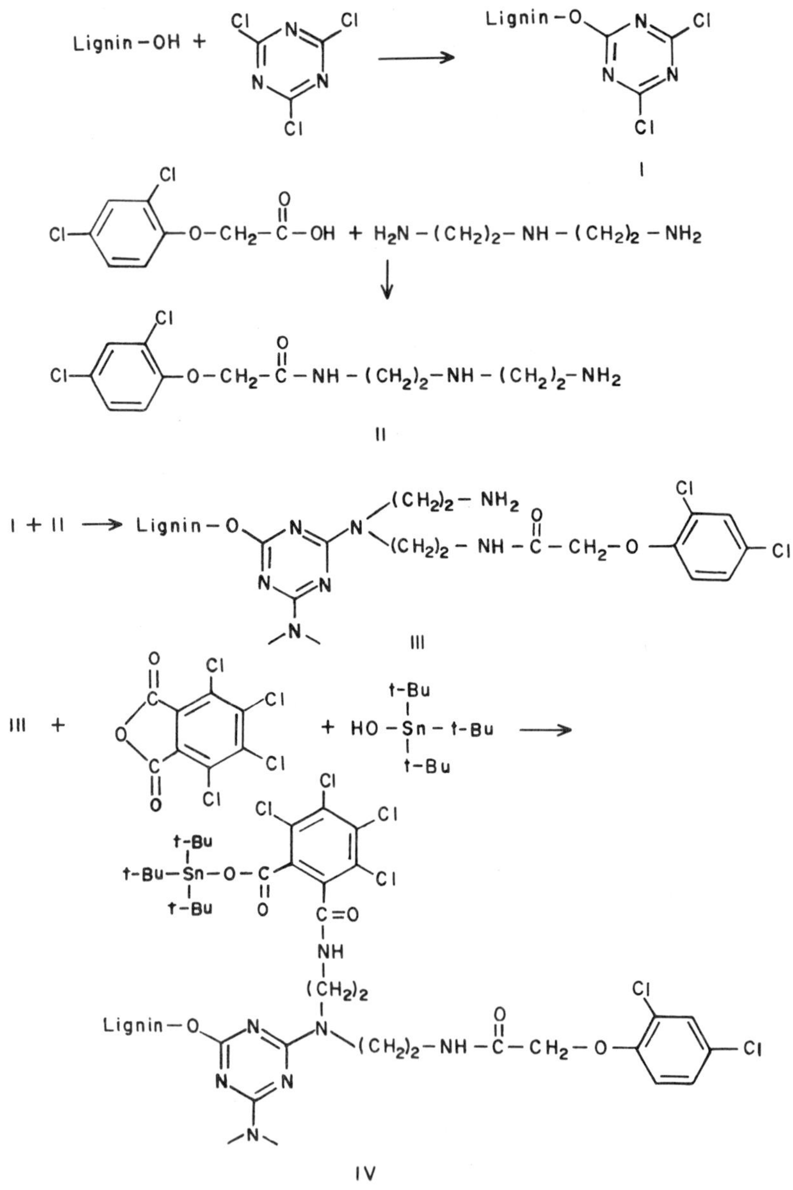

SCHEME-9

attached to kraft lignin. The product is capable of undergoing hydrolysis to re-lease a potential equivalent of octamethyl pyrophosphoramide..

It was reported that the reaction of phosphoryl chloride with water and kraft lignin is smooth and gives a product after capping with dimethylamine. The idealized representation of the reactions described above is shown in Scheme 10. Efforts have been made to prepare lignin-phosphorus derivatives containing phosphate, thiophosphate, pyrophosphate, and dithiopyrophosphate groups. The sodium salts of thiolignin and cellulose were condensed with dialkyl phos-phochloridothioate to afford pesticide–polymer systems. Some carboxymethyl ethers of kraft lignin have insecticidal and fungicidal activity. A hydrolyzable lignin ether of 4-hydroxy-2,3,5-trichloropyridine was prepared by Allan from lignin and 2,3,4,5-tetrachloropyridine.

Naruse and co-workers[39] have reported a series of copolymers with dithio-carbamate, isothiocynate, and isothiourea bonds. These copolymers decompose physically or chemically to release sulfur and sulfur-containing compounds that show herbicidal and fungicidal activity. A series of polyethyleneimine s-alkyldi-thiocarbamates were prepared from polyethyleneimine, sodium dithiocarbamate, and alkyl chlorides, aralkyl chlorides, substituted alkyl esters of chloroacetic acid, β-chloroethyl esters of carbamic acid, carboxylic acid, and N-substituted chlorocarboxamides. Polyethyleneimine sodium thiocarbamate was prepared by the action of sodium hydroxide and carbon disulfide on polyethyleneimine. The bioactive groups are covalently linked to the polymers (Scheme 11).

A series of poly(s-vinyl alkyldithiocarbamates) were prepared from polyvinyl chloride and sodium alkyl dithiocarbamates. These polymers show herbicidal activity. They inhibited the growth of barnyard grass and radish seedlings. They were observed to decompose in aqueous alkaline solution to give thiocyanate and other degradation products. Decomposition in acid medium gave low-molecular-weight compounds.

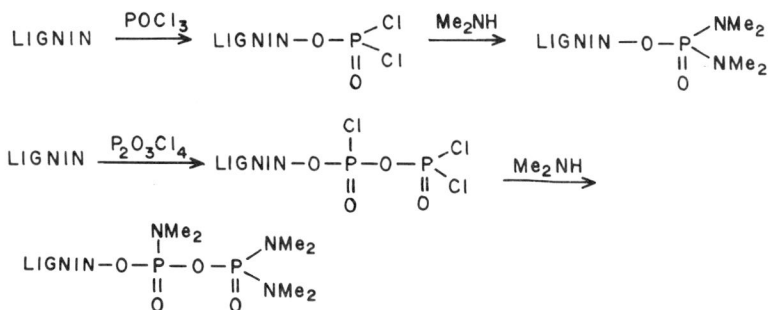

SCHEME 10

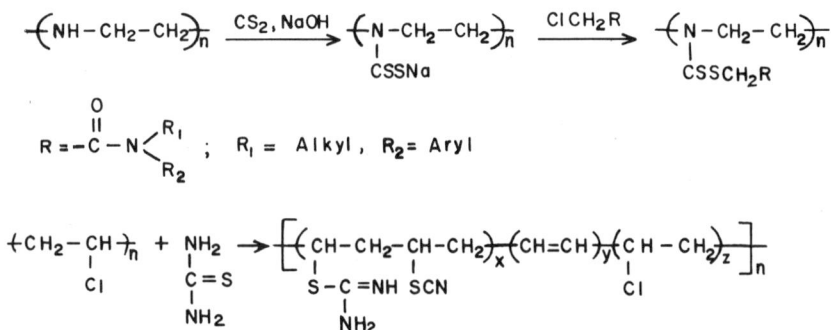

$$\left(\!NH\!-\!CH_2\!-\!CH_2\!\right)_n \xrightarrow{CS_2,\,NaOH} \left(\!N\!-\!CH_2\!-\!CH_2\!\right)_n \xrightarrow{ClCH_2R} \left(\!N\!-\!CH_2\!-\!CH_2\!\right)_n$$
$$\underset{CSSNa}{} \qquad\qquad \underset{CSSCH_2R}{}$$

$$R = -\overset{O}{\underset{}{C}}-N\!\!\begin{array}{c} R_1 \\ R_2 \end{array} \quad ; \quad R_1 = Alkyl, \quad R_2 = Aryl$$

$$\left(\!CH_2\!-\!CH\!\right)_n + \underset{\underset{NH_2}{C=S}}{NH_2} \longrightarrow \left[\left(\!CH\!-\!CH_2\!-\!CH\!-\!CH_2\!\right)_x\!\!\left(\!CH\!=\!CH\!\right)_y\!\!\left(\!CH\!-\!CH_2\!\right)_z\right]_n$$
$$\underset{Cl}{} \qquad\qquad\qquad \underset{\underset{NH_2}{S-C=NH}}{}\ \underset{}{SCN} \qquad\qquad \underset{Cl}{}$$

SCHEME II

A totally new class of polymers, which are decomposed by water and show controlled-release bioactivity, was introduced by Beasley and Collins.[40] These polymers were obtained by the addition of a carboxylic acid to a metal ion in the presence of an aldehyde catalyst. They decompose slowly to release the original carboxylic acid. The release rate has been found to vary from days to years. The acids that have been polymerized include the herbicides 2,4-dichloro-phenoxyacetic acid, 2,4,5-trichlorophenoxyacetic acid, 2-(2,4,5-trichlorophen-oxy propionic) acid, 2-methoxy-3,6-dichlorobenzoic acid, and 4-amino-3,5, 6-trichloropicolinic acid. The metals that can be used are iron, cobalt, nickel, titanium, manganese, and chromium. A typical example is shown in Scheme 12. On heating, the rate of degradation of these polymers decreases while their molecular weight and hardness increase. The rate is also controlled by varying the surface-to-volume ratio. The polymer prepared from 2,4-D and iron has shown exceptional weed control in field trials for a period up to one year.

2.5. Derivatization

2.5.1. Esters

A. Steroids. 1. Testosterone. Among the steroids, testosterone, nortestos-terone, estrone, progesterone, and adrenocortical hormones were converted to long-acting forms by conversion to esters. Miescher et al.[41] prepared a series of testosterone aliphatic esters. The duration of activity as a function of chain length was determined for 11 testosterone esters. It was shown that the initiation of action and intensity are greatest for formate, acetate, and propionate. The solubility at the injection site, which favors easy absorption and facile hy-drolysis, appears to be responsible for the rapid onset of action and improved

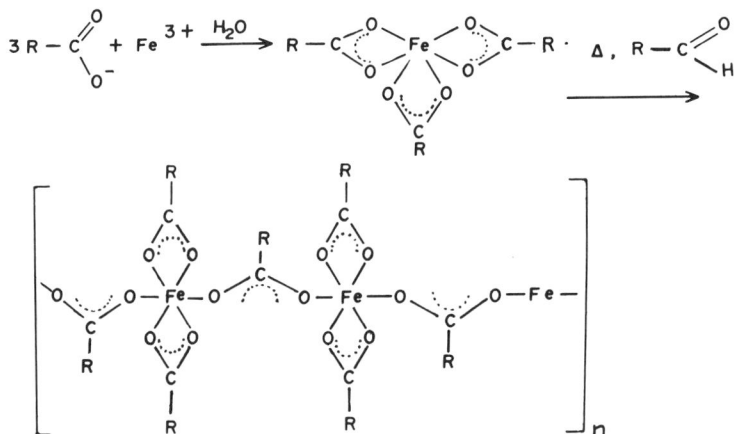

SCHEME 12

activity. Butyrate, isobutyrate, and valerate exhibit lower duration and intensity, probably due to decreased solubility and slower enzymatic hydrolysis. The differences in the initiation, intensity, and duration of activity are significant for short-chain and sterically unhindered esters. Among the various cycloaklyl esters of testosterone, testosterone β-cyclopentyl propionate has been shown to have the maximum anabolic and androgenic activity and duration of response.[42] The propionate ester is a good example to illustrate the superiority of a chemical sustained-release derivative of a drug over a physical formulation of testosterone.

The effects of chain branching of testosterone esters on the activity and duration through ester hydrolysis inhibition have been reported.[43] Esters with dialkyl-substituted chains show poor activity, with the exception of testosterone isobutyrate. Substituted cyclohexane carboxylic acid esters of testosterone have interesting structure–activity relationships. The activities of 3- and 4-alkyl-substituted cyclohexane carboxylates are similar to those of the heptanoate and nonanoate straight-chain acid esters. Chain extension by including methyl cyclohexyl acetate resulted in lower potency and shorter duration. Phenoxyalkanoate esters of testosterone have high potency and duration of activity.[44]

2. *Nortestosterone.* Like testosterone, nortestosterone has androgenic and anabolic properties. Phenylpropionate, decanoate, decosanoate, and oleate have been prepared.[45] These derivatives possess good duration of activity and better anabolic/androgenic ratios. Many 19-nortestosterone terpenoate esters were evaluated for anabolic activity.[46]

3. *Estradiol/Estrone.* Long-acting estrogens have been prepared via the prodrug ester derivatives. The early examples are mono- and bisesters of estradiol

and estrone.[47] Estrone esters possess specific structure–activity correlation based on length of the ester chain. Rapid decrease in duration is observed in esters larger than the octanoate, probably due to the effect of aqueous solubility, rate of absorption, and bioavailability.

4. *Hydrocortisone.* Among the adrenocortical hormones, the antiinflammatory agent hydrocortisone has been converted to hydrocortisone-21-caproate.[48] It showed prolonged liver glycogen activity. Hydrocortisone trimethylacetate is more readily absorbed but is not very potent compared to the parent drug. Cortisone-21-trimethyl acetate was found to be longer acting than cortisone. This is attributed to a combination of decreased rate of absorption and ester hydrolysis. Hydrocortisone-17-butyrate and 17-valerate esters showed satisfactory potency with topical application. Whether these esters act as prodrugs has not been conclusively demonstrated. The activity of betamethasone-17α-20-orthoesters, is believed to be due to their hydrolysis by acidic components present in sweat.[49] Difluorocorticosteroid 17,21-methyl orthoesters, 17-monoesters, and 17,21-diesters have been prepared and shown to have prolonged activity.[50]

B. Neuroleptics. These drugs are used in the treatment of mental disorders, especially psychoses. The tranquilizers are prescribed to cure common psychoneuroses and in somatic disorders. A common method to prolong neuroleptic action is conversion to ester or salts. The esters have high oil/water partition coefficients and thus enable slow drug diffusion from the oil vehicle to the tissue fluid at the injection site. After the hydrolysis by esterase, the hydrolyzed parent drug is transported from the injection site.[51] Like steroids, the longer the fatty acid ester chain, the greater the duration of activity. If the aqueous solubility of a salt derivative is low, the duration of effect will be long. Long-acting neuroleptics have been prepared by derivatizing phenothiazine, thioxanthene, dibenzazepine and diphenyl butyl piperidine. Perphenazine enanthate [heptanoic acid-2-(4-[3-(2-chlorophenothiazin-10-yl)propyl]-1-piperazinyl)ethyl ester] (I, Scheme 13) has been prepared as an injectable long-acting neuroleptic.[52] The undecylenate (II) and palmitate (III) of pipothiazine (2-diemthylsulfamoyl-10-[3-(4-hydroxyethyl-piperidino)propyl] phenothiazine) have been prepared and evaluated for sustained neuroleptic action. Fluphenazine, a potent tranquilizer, is a derivative of phenothiazine. Fluphenazine heptanoate (IV) and decanoate (V) esters showed sustained action. An evaluation of all these esters showed that the duration of effect increased as a function of the increase in the chain length of the ester.[53] Flupenthixol [2-trifluoromethyl-(3-[1-(2-hydroxyethyl)-4-piperazinyl] propylidene)-9-thioxanthene] (VI) is an orally effective thioxanthene neuroleptic used in psychiatry. Frequent dosing to maintain therapeutic levels leads to side effects. These shortcomings have been eliminated by changing the route of administration and chemical form administered. The decanoate (VII) ester and intramuscular administration have met with wide acceptance in psychiatry.

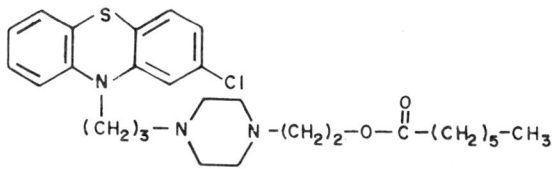

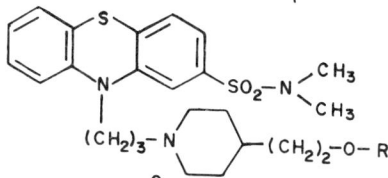

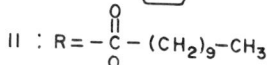

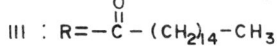

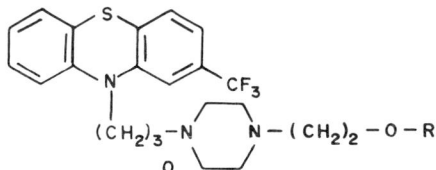

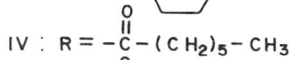

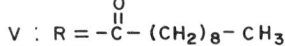

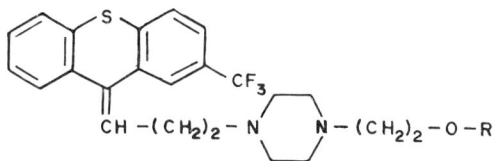

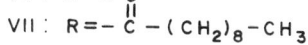

SCHEME 13

C. Dibenzazepines. In the treatment of endogenous and reactive depression, imipramine (5-[3-diemthylaminopropyl]-10,11-dihydro-5H-dibenz[b,f]azepine) (VIII) is used as an antidepressant. As a sustained dosage form, its pamoate salt has been prepared for oral administration (IX) (Scheme 14). Wilson et al.[54] reported that the pamoate salt showed reduced side effects.

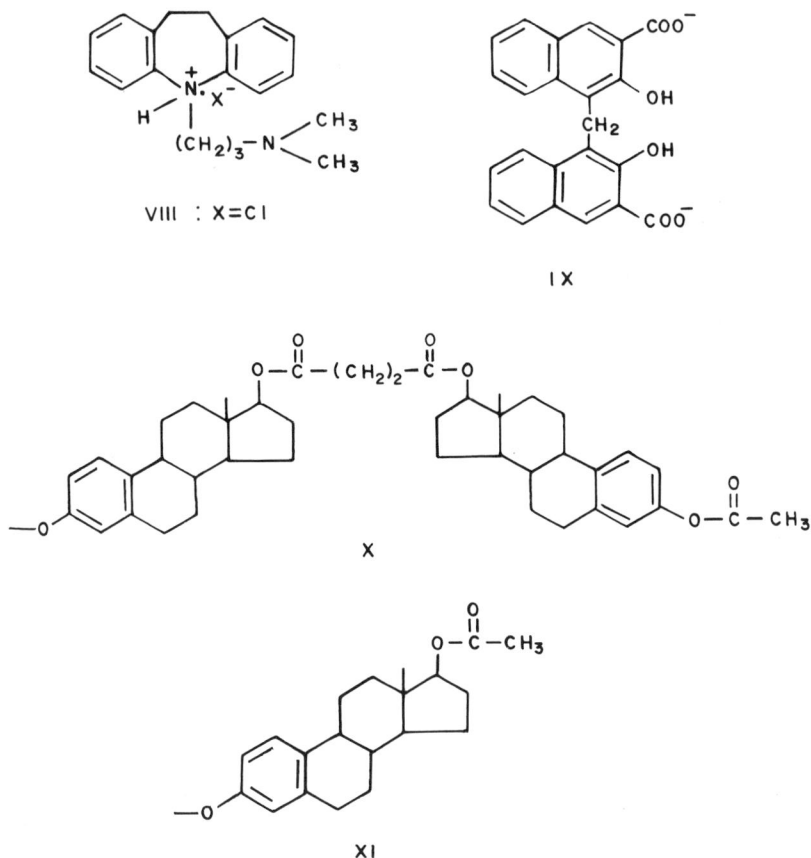

VIII : X=Cl

IX

X

XI

SCHEME 14

D. Local Anesthetics. Polyethylene glycol carbamate derivates of procaine have been prepared. They showed that the sustained action depends on the type of polyethylene glycol used.[55] Macromolecular dextran-procaine derivatives exhibit sustained anesthesia.[56] Procaine has been converted to sterically hindered amides, commonly called xylidides. Their duration of activity is due to the resistance of the amide bond to enzymatic hydrolysis. Substitution by methyl groups at positions *ortho* to the amide bond enhances resistance to hydrolysis. Amphetamine, ephedrine, dopamine, dapsone, isoniazid, antimalarials, anti-hypertensives, and hypoglycemics have been converted to their esters.

A novel approach to prolong estrogenic activity is to synthesize oligomeric estradiol esters.[57] These oligomers contain two to four steroid molecules covalently linked via succinate ester (X). They are prepared by a new appraoch. The reaction of 3-acetoxyestradiol-17β-hemisuccinate with N,N'-carbonyldimidazole gave an intermediate that was then condensed with estradiol or its acetate ester to produce a variety of oligomers.[58] Many of these have enhanced duration of estrogenic activity. Increasing the chain length leads to increased duration of vaginal estrus up to 120 days. Dimeric derivatives of isomeric esters such as XI also exhibit prolonged activity (Scheme 14).

2.5.2. Ethers

Several steroids with hydroxyl groups have been converted to their ethers. Cross et al.[59] reported the synthesis of many 2'-tetrahydroxypyranyl ethers of nortestosterone. They have also prepared 17β-tetrahydropyranyl ethers of androstane and 19-norandrostane. On oral administration, these acid-labile ethers exhibit satisfactory bioactivity, due to acid hydrolysis in the stomach. With subcutaneous injection, they showed low activity. An unusual delay in reaching the peak response level was reported with 17-trimethylsilyl ether of testosterone.[60] This ether derivative is more potent than the propionate ester. Many 17α-alkynyl estradiol-3-cyclopentyl ether and 17-benzoate esters show prolonged antigonadotrophic activity twice that of the parent ovarian inhibitor quinestrol (17α-ethynyl estradiol-3-cyclopentyl ether.[61]

Meli and Steinetz[62] have reviewed steroidal biology and metabolism and highlighted the factors that determine their bioactivity. Steroidal 3-ethers and 3-enol ethers are bioactive. The sustained activity is due to slow hydrolysis of the ether linkage. The hydrolytic products are different from the parent steroid. The ethers have altered pharmacokinetic properties. Their adsorption, distribution, storage, metabolism and excretion have significant effects on the duration of activity. This has been demonstrated in the case of 3-enol ethers of methyl testosterone and 17αacetoxyprogesterone[63]. The lowered activity of enol ethers administered subcutaneously may be due to poor absorption and storage in fatty tissues rather than to slow hydrolysis. Bioactivity may be decreased due to lack of metabolic inhibition. The reduced bioactivity of 19-norsteroid enol ethers administered orally may be due to the absence of the 19-methyl group, which would inhibit their rate of metabolism.[64]

Most of the esters act as prodrugs and require hydrolysis to the parent steroid to exhibit bioactivity. With steroidal ethers, their pharmacokinetics and bioavailability govern activity and duration. Many estrogen-3-ethers exhibit less activity when they are introduced subcutaneously rather than orally[65]. Ethynyl estradiol-3-cyclopentyl ether administered orally accumulates in body fat and is released slowly. The alkyl groups on C-17 in natural and synthetic estrogens were implicated in their heptatoxicity. Many estradiol-3-esters-17-ethers were synthesized to overcome their toxicity. Estradiol-17-cyclohexyl ether 3-propionate and estradiol-17-cyclo-octenyl-ether 3-benzoate showed a separation of estrogenic and hepatotoxic effects.

2.6. Analogs

Another approach to enhance the duration of drug effects is synthesis of drug analogs. Haloperidol (4-[4-chlorophenyl-4-hydroxypiperidinol]-4'-fluorobutyro-phenone) and pimozide [1-(1-[4,4-bis(p-fluorophenyl)butyl]-4-piperidyl)-2-benzimidazolinone] (XIII) are two orally active central nervous system depressants, sedatives, and tranquilizers (Scheme 15). A comparative pharmacological study showed that pimozide is much less toxic and long-acting than haloperidol. Penfluridol (XIII) (4-[4-chloro-α,α,α-trifluoro-m-tolyl]-1-[4,4-bis(p-fluorophenyl)-butyl]-4-piperidinol),[51] an analog of pimozide, has a duration of activity of one week.[66] Initiation of activity is smooth and gradual, and peak activity is attained in 24–48 h. The toxic effects are less than those of other neuroleptic agents. Chemically combining the phenothiazine and benzimidazoline moieties into one molecule produced many nontoxic neuroleptic agents of long duration of activity.

Many semisynthetic penicillins that are analogs of 6-aminopenicillanic acid have different potency, spectrum of activity, toxicity, and duration of activity. α-Phenoxyalkyl penicillin analogs are a good example. Among the cephalosporin analogs, cefazoline (XIV),[51] cephaloridine (XV),[52] and cephalothin (XVI)[53] show significant differences in the duration of activity.

The use of long-acting forms of sulfanilamides is relatively new. Sulfamethoxy-pyrazine (XVII),[54] sulfadimethoxine (XVIII),[55] sulfamethoxydiazine (XIX),[56] and sulfamethoxypyridazine (XX)[57] are designed as long-acting sulfanilamides (Scheme 15).

3. RELEASE MECHANISMS

In some conventional controlled-release formulations, the active ingredient is adsorbed on inert materials such as silica gel, mica, or activated charcoal. The active agent in such systems is released by the process of desorption, and it is difficult to get the desired release rates of the pesticide. Controlled-release systems are designed with a view to obtain better control over the release rates. In such systems the mechanism of release can be diffusion through a rate-controlling membrane, erosion of a degradable material, or a retrograde chemical reaction. In reservoir systems where the pesticide is dissolved, trapped, or microencapsulated in a polymeric matrix, it is released by diffusion through the polymer, when the pesticide is chemically bonded to a polymer the bond has to be cleaved by environmental agents and the release mechanism depends on a retrograde chemical reaction.

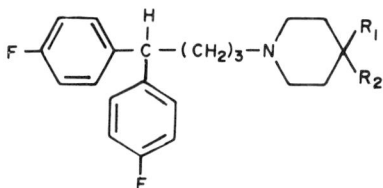

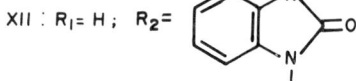

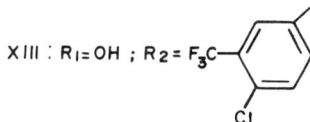

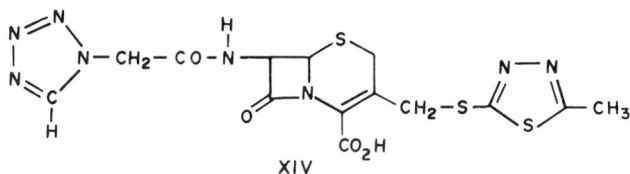

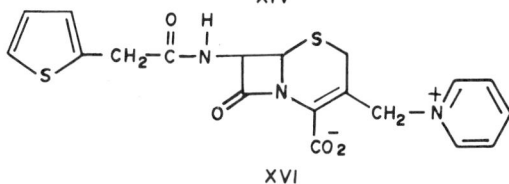

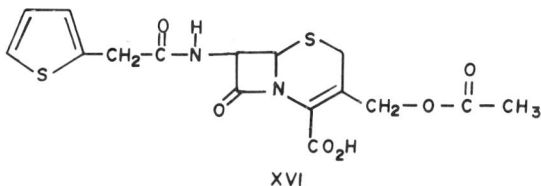

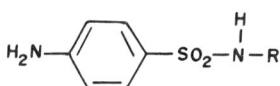

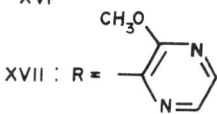

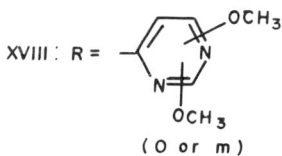

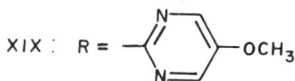

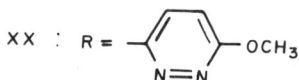

XII : $R_1 = H$; $R_2 =$

XIII : $R_1 = OH$; $R_2 = F_3C$

XIV

XV

XVI

H_2N—SO_2—$\overset{H}{N}$—R

XVII : R =

XVIII : R =

(O or m)

XIX : R =

XX : R =

SCHEME : 15

The rate of cleaveage of definite and identifiable chemical bonds depends on the type of bond, the nature of the macromolecule, and the properties of the environment. For example, anhydride bonds are hydrolyzed faster than ester or amide linkages. Neighboring groups in the polymeric backbone also influence the release rates. Hydrophilic moities tend to increase the rate of hydrolytic bond cleavage, while hydrophobic groups retard the bond fission. Linear and amorphous polymers undergo faster hydrolysis than cross-linked and crystalline polymers. The rate of pesticide release by retrograde chemical reaction may also vary depending on whether the hydrolysis reaction is taking place on the surface of an insoluble particle or in solution.

The hydrolysis can be homogenous, heterogenous, or a combination of both. The release rate depends on the reaction kinetics, the rate of diffusion of the active agent through the polymer, and boundary layer effects.[67,68] A water-soluble polymer undergoing homogenous hydrolysis with no boundary layer effects follows first-order kinetics, and the reaction rate limits the release rate. Such systems are ideal systems.

If the reaction is heterogenous, the rate of release is also governed by the geometry and size of the pesticide–polymer combination. Small particles have high surface-to-volume ratios, and hence the smaller the particle the faster the release rates. Neogi and Allan[69] have derived the following rate expression for a heterogenous reaction on the surface of insoluble spherical particles:

$$\frac{d_M}{dt} = -\rho n 4\pi r^2 \frac{dr}{dt} = nk 4\pi r^2 C_0$$

where d_M/dt is the rate of release of pesticides, n is the number of particles of average radius r at the time t, k is the rate of constant for the hydrolysis reaction, ρ is the density, and C_0 is the concentration of pesticide–polymer linkages. If the hydrolyzed polymer is water-soluble, C_0 is considered constant, because as one pesticide molecule is removed from the surface, another pesticide–polymer bond comes in contact with water.

Allan and Neogi have also derived the following expressions for the rate of release when the hydrolyzed polymer is water insoluble.

$$-\frac{dc}{dt} = K_2 C$$

$$-\frac{dc}{c} = K_2 dt$$

$$\frac{\ln C_0}{c} = K_2 t$$

where c is the concentration of pesticide per unit weight at time t, and K_2 is the degradation rate constant.

Baker and Lonsdale[67] have discussed a theoretical example where heterogenous or homogenous reactions occur without any diffusion of the free released pesticide.

The release rates depend on the degree of substitution, the pH of the hydrolysis medium, and the geometry, microstructure, and size of the system. In the case of natural polymers, the pattern of herbicide release has been explained on the basis of microstructure and unequal reactivities of the hydroxyl groups in each anhydroglucose unit toward esterification. α-Cellulose fiber contains 50-70% dispersed crystalline and 30-50% amorphous regions.[70] The esterification does not take place uniformly. The amorphous regions are preferentially esterified in nonpolar, nonswelling reaction media. From tosylation studies, the relative reactivities of the C_6, C_2, and C_3 hydroxy groups have been reported to be in the ratio $215:33:1$, respectively.[71] Kinetic studies have also shown that the primary hydroxy group reacts 58 times faster than the adjacent secondary hydroxy groups.[72] Hence, the density and pattern of substitution varies in the α-cellulose backbone.

As the degree of substitution is increased, the hydrophilic character of the herbicide–polymer combination decreases. The hydrophilicity can become so small that water cannot effectively permeate the system and trigger the hydrolytic release of the pesticide.[24] This explains the observation that Douglas fir bark or sawdust containing greater quantities of 2,4-D release less active ingredient than the corresponding preparations with lower amounts of covalently linked herbicide.[73] The release rates of 4-(2,4-dichlorophenoxy)butyric acid from the bark of western red alder, Douglas fir, and hemlock were reported to be in the ratio $1:2:3$. This can be attributed to the difference in the intrinsic chemistry and morphology of these natural polymers. Thus it is evident that the nature of the polymeric substrate can also modify the rate of release.

Homopolymers in which the pesticide molecule is bonded directly to a polymeric backbone exhibit extremely slow rates of hydrolysis. It was demonstrated that increasing the distance between the pesticide and the main chain by extending the pendent chain length would enhance the rate of ester hydrolysis and hence the rate of pesticide release. The hydrolysis of pendent ester groups has been reported to be enhanced by incorporating a carboxyl group in the backbone. Similar hydrophilic neighboring group participation in the hydrolysis of pesticides linked to polymers through ester linkages has been reported. The rate of hydrolysis of many copolymers containing different percentages of methacrylic acid has been investigated.

The neighboring carboxylic group of the polyacrylic acid chain is believed to be responsible for rapid hydrolysis of ester groups.[74] A six-membered acid anhydride intermediate is assumed to be formed due to the thermal nucleophilic attack of neighboring carboxylate ion on the carbonyl carbon of the ester group. The formation of the six-membered acid anhydride ring has been shown to be the rate-determining step. To exhibit herbicidal activity it should undergo further hydrolysis to give 2,4-D.

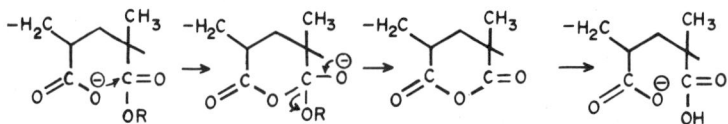

An alternative explanation to rationalize the observed enhancement of the rate of ester hydrolysis is that even though the ester and carboxyl groups are not in the vicinal positions, but are located several atoms away from each other in the polymer chain, they can come nearer to each other due to the flexibility and mobility of the polymer chain. Moreover, they are not sterically hindered, which also favors hydrolysis.

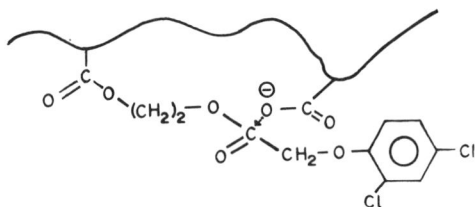

It has been demonstrated that the formation of a six-membered cyclic anhydride intermediate results in considerable steric hindrance between the two axial substituents, which may inhibit the rate of hydrolysis. The possible structures of the six-membered acid anhydride ring intermediate are shown below:

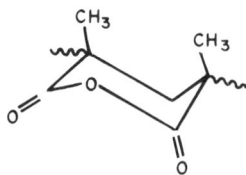

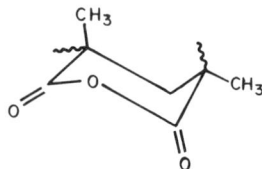

ISOTACTIC   SYNDIOTACTIC

Since hydroxyl groups are considerably less hydrophilic than carboxyl groups, they could be expected to be less effective in catalyzing ester hydrolysis. The rates of hydrolysis of copolymers of 2-acryloyloxyethyl 2,4-dichlorophenoxy-acetate and 2-hydroxyethyl methacrylate in varying proportions have been reported. It was found that if the amount of hydrophilic component in the monomers is not sufficient, the rates will not be appreciable.

Swelling of polymers has been observed to be accompanied by autoaccelera-tion of the hydrolysis rate. Such autoacceleration in the rate of hydrolysis occurs only beyond a certain limiting degree of hydrophilicity. Branched copolymers with molecular weights in the range 12,000–15,000 show rates of hydrolysis that increase as the hydrolysis progresses. Many branched polymers also slowly swell and ultimately dissolve. Hydrolysis of a copolymer need not depend on its degree of branching or its molecular wieght. Cross-linked polymers hydrolyze at a constant rate. They do not generally go into solution even after swelling.

Studies on the effect of the aminimide group on ester hydrolysis has been reported based on the studies on 2-acryloyloxyethyl-2,4-dichlorophenoxyacetate and trimethylamine methacrylamide. The rate of hydrolysis showed a preliminary increase that remained relatively constant. The aminimide group appears to assist the hydrolysis of the neighboring ester linkage in a fashion similar to that of the carboxyl group. Copolymers containing aminimide groups hydrolyze more slowly than copolymers with carboxyl groups under similar conditions.

The rates of hydrolysis of metribuzin homopolymers having amide, urea, and carbamate bonds in water at 25°C have been reported. Polymers prepared using acrylic and methacrylic monomers exhibit similar rates of release.[75] Satisfactory rates of hydrolysis were obtained due to the electronic nature of the triazine ring and the presence of amide and urea bonds. Over an extended period, these polymers depolymerize to give a mixture of the monomer and polymer. The de-polymerization has been ascribed to the low ceiling temperature of methacrylic polymers. The rate of metribuzin release did not diminish during the first seven days. As the amide bonds are hydrolyzed, polyacrylic acid units are formed, which may increase the hydrophilicity, and catalysis by the carboxylic group results in enhanced release rates.

The role of structural features of the polymer in determining the release rates of pendent substituents was studied with many natural and synthetic polymers containing metribuzin as a pendent substituent.[76] It was shown that substitu-tion of herbicide in the polymeric chain reduces the intermolecular hydrogen bonding in polymers and makes them soluble in organic solvents. The degree of o-substitution in such polymers changes the hydrophilicity of the copolymers. In water-insoluble copolymers, the release rates in water were found to decrease with time. As hydrolysis takes place on the surface, hydroxy groups are liberated, resulting in more hydrogen bonding. This prevents the entry of water through the outer surface and reduces the rate of hydrolysis.

REFERENCES

1. A. S. Crafts, *The Chemistry and Mode of Action of Herbicides*, Interscience, New York, 1961, p. 217.

2. G. Faerber, Brit. Patent 826,831 (1960).

3. E. Baltazzi, U.S. Patent 3,343,941 (1967).

4. G. G. Allan, J. W. Beer, M. J. Cousin, W. J. McConnell, J. C. Powell, and A. Yahiaoui, "Polymeric Drugs for Plants," in L. G. Donaruma and O. Vogel, Eds., *Polymeric Drugs*, Academic, New York, 1978, p. 193.

4a. K. G. Das, *Proc. 5th Int. Symp. Controlled Release Bioactive Materials, Gaithersburg, MD*, 1978, p. 3.29.

5. G. G. Allan, J. W. Beer, M. J. Cousin, and R. A. Mikels, in A. F. Kydonieus, Ed., *Controlled Release Technologies: Methods, Theory and Applications*, Vol. II, CRC Press, Florida, 1980, pp. 7–62.

6. M. J. Cousin, "The Controlled Release of Herbicides from Biodegradable Substrates," Ph.D. Thesis, Univ. Washington, Seattle, 1978.

7. R. del Moral and C. H. Muller, Alleopathic Effects of *Eucalyptus camaldulensis, Am. Midl. Nat.*, 83, 254 (1970).

8. H. D. Jakubka and E. Busch, *Z. Chem.*, 13, 105 (1973).

9. J. F. Friedhoff, "The Potential of Controlled Release Herbicides for the Suppression of Unwanted Vegetation," Ph.D. Thesis, Univ. Washington, Seattle, 1975.

10. K. Akagane and G. G. Allan, "Long-Term Protection V, Mutual Interaction of Active Toxicants in Antifouling Polymers Consisting of Multicomponent Toxicants," *Shikizai Kyokaishi*, 46, 437 (1973).

11. K. Akagane and K. Matsura, "Long-Term Protection I, Effects of Toxic Polymers on Long-Term Protection," *Shikizai Kyokaishi*, 45, 691 (1972).

12. Z. M. O. Rzaev, *Chem. Tech.*, 1, 58 (1979).

13. T. St. Pierre and E. Schacht, in D. H. Lewis, Ed., *Proc. 7th Int. Symp. Controlled Release Bioactive Materials, Ft. Lauderdale, FL*, 1980, p. 180.

14. G. G. Allan, J. R. Fox, and N. A. Kong, "Critical Evaluation of the Potential Sources of Chitin and Chitosan," in *Proc. 1st Int. Conf. Chitin Chitosan, Boston*, M.I.T., Sea Grant Program, 1978.

15. H. Naruse and K. Maekawa, *J. Fac. Agric. Kyushu Univ.*, 21, 153 (1977).

16. H. Naruse and K. Maekawa, *J. Fac. Agric. Kyushu Univ.*, 21, 107 (1977).

17. G. G. Allan and D. D. Halabisky, *J. Appl. Chem. Biotechnol.*, 21, 190 (1971).

18. K. Akagane and G. G. Allan, *Shikizai Kyokaishi*, 45, 479 (1972).

19. C. L. McCormick, K. E. Savage, and B. Hutchinson, in R. L. Goulding, Ed., *Proc. Int. Controlled Release Pesticide symp., Oregon*, 1977, p. 28.

20. S. Bittner, I. Perry, and Y. Knobler, *Phytochem.* 16, 305 (1977).

21. V. V. Dovlatyan and D. A. Kostanyan, *Izv. Akad. Nauk. Arm. SSR Khim. Nauki*, 18, 325 (1965).

22. R. K. Vander Meer and C. S. Lofgren, in D. H. Lewis, Ed., *Proc. 7th Int. Symp. Controlled Release Bioactive Materials, Ft. Lauderdale, FL*, 1980, p. 164.

23. F. W. Harris and L. K. Post, *J. Polym. Sci. Part A-1*, 13, 225 (1975).

24. A. N. Neogi, "Polymer Selection for Controlled Release Pesticides," Ph.D. thesis, Univ. Washington, Seattle, 1970.

25. R. M. Wilkins, "The Design of Polymers for the Sustained Release of Selected Herbicides," M.S. thesis, Univ. Washington, Seattle, 1969.

26. F. W. Harris, W. A. Feld, and B. Bowen, in F. W. Harris, Ed., *Proc. Int. Controlled Release Pesticide Symp.* 1975, p. 334.

27. C. L. McCormick and M. M. Fooladi, in R. W. Baker, Ed., *Proc. 6th Int. Symp. Controlled Release Bioactive Materials*, 1979, III 27.

28. F. W. Harris, M. R. Dykes, J. A. Baker, and A. E. Aulabaugh, in H. B. Scher, Ed., *Controlled Release Pesticides (ACS Symp. Ser. 53)*, Chem. Soc., Washington, DC, 1977, p. 102.

29. F. W. Harris and C. O. Arah, *Polym. Preprints,* **21**, 107 (1980).

30. F. W. Harris and C. O. Arah, in D. H. Lewis, Ed., *Proc. 7th Int. Symp. Controlled Release Bioactive Materials, Florida*, 1980, p. 172.

31. N. A. Ghanem, N. N. Messiha, N. E. Ikladious, and A. F. Shaaban, in D. H. Lewis, Ed., *Proc. 7th Int. Symp. Controlled Release Bioactive Materials, Florida*, 1980, p. 160.

32. V. J. Castelli and W. L. Yeager, in D. R. Paul and F. W. Harris, Eds., *Controlled Release Polymeric Formulations (ACS Symp. Ser.* **33**), Am. Chem. Soc., Washington, DC, 1976, p. 233.

33. W. L. Yeager and V. J. Castelli, "Antifouling Applications of Various Tin-Containing Organometallic Polymers," *173rd Nat. Am. Chem. Soc. Meet., New Orleans*, March 1977.

34. V. J. Castelli, D. M. Anderson, C. E. Mullin, and W. L. Yeager, "Polymers for Antifouling Drag-reducing Coating Systems," *Rep. MAT-76-20*, David W. Taylor Naval Ship Res. Develop. Center, Anapolis, MD, May 1976.

35. G. G. Allan and A. N. Neogi, U.S. Patent 4062,855 (1977).

36. K. Akagane and G. G. Allan, *Shikizai Kyokaishi,* **46**, 622 (1973).

37. K. Akagane, Jap. Patent 74,125,521 (1974).

38. K. Akagane, G. G. Allan, and E. M. Passot, *Pap. Puu*, **56**, 1 (1974).

39. H. Naruse and K. Maekawa, *J. Fac. Agric. Kyushu. Univ.,* **21**, 173 (1977).

40. M. L. Beasley and R. L. Collins, *Science*, **169**, 769 (1970).

41. K. Miescher, A. Weltstein, and E. Tschopp, *Biochem. J.,* **30**, 1977 (1936).

42. W. Sakamoto, G. S. Gordan, and E. Eisenberg, *Proc. Soc. Exp. Biol. Med.,* **76**, 406 (1951).

43. D. Gould, L. Finckenor, E. B. Hershberg, J. Cassidy, and P. L. Perlman, *J. Am. Chem. Soc.,* **79**, 4472 (1957).

44. D. Gould, L. Finckenor, P. Perlman, J. Cassidy, S. Margolin, M. T. Spoerlein, and E. B. Hershberg, *Chem. Ind.,* **1955**, 1424.

45. J. de Visser and G. A. Overbeck, *Acta Endocrinol.,* **35**, 405 (1960).

46. G. Pala, S. Casadio, A. Mantegani, G. Bonardi, and G. Coppi, *J. Med. Chem.,* **15**, 955 (1972).

47. K. Miescher, C. Scholz, and E. Tschopp, *Biochem. J.,* **32**, 725 (1938).

48. C. A. Winter and C. C. Porter, *J. Amer. Pharm. Ass. (Sci. Ed.),* **46**, 515 (1957).

49. A. W. McKenzie and R. M. Atkinson, *Arch. Dermatol.,* 89, 741 (1964).

50. R. Gardi, R. Vitali, G. Falconi, and A. Ercoli, *J. Med. Chem.*, **15**, 556 (1972).

51. A. G. Ebert and S. M. Hess, *J. Pharmacol. Exp. Ther.*, **148**, 412 (1965).

52. G. M. J. Van Kempen, *Psychopharmacol.*, **21**, 283 (1971).

53. L. Julou, G. Bourat, R. Ducrot, J. Fourmel, and C. Garret, *Acta Psychiat. Scand.*, 49(Suppl. 241), 9 (1973).

54. I. C. Wilson, A. M. Rabon, H. A. Merrick, A. E. Knose, J. P. Taylor, and W. J. Buffaloe, *Psychosomatics*, **7**, 251 (1966).

55. B-Z. Weiner and A. Zilkha, *J. Med. Chem.*, **16**, 573 (1973).

56. L. Molteni and F. Schrollini, *J. Med. Chem.*, **9**, 618 (1974).

57. H. Kuhl and H. D. Taubert, *Steroid*, **22**, 73 (1973).

58. R. W. Talley, C. R. Haines, M. N. Waters, I. S. Goldenberg, K. B. Olson, and H. F. Bisel, *Cancer*, **32**, 315 (1973).

59. A. D. Cross, I. T. Harrison, P. Crabbe, F. A. Kinel, and R. I. Dorfman, *Steroids*, **4**, 229 (1964).

60. F. J. Saunders, *Proc. Soc. Exp. Biol. Med.*, **123**, 303 (1966).

61. R. Gardi, R. Vitali, and G. Falconi, U.S. Patent 3,840,568 (Oct. 1974).

62. A. Meli and B. Steinetz, *Trans. N.Y. Acad. Sci.*, **28**, 623 (1965–1966).

63. G. Falconi, R. Gardi, G. Bruni, and A. Ercoli, *Endrocrinol.*, **69**, 638 (1961).

64. A. Meli, *Endocrinol.*, **72**, 715 (1963).

65. A. Ercoli, *Hormonal Steroids*, **1**, 17 (1964).

66. F. A. J. Janssen, C. J. E. Niemegears, K. H. L. Schellekens, F. M. Lenaerts, F. J. Verbrugger, J. M. Van Nultan, and W. K. A. Schaper, *Europ. J. Pharmacol.*, **11**, 139 (1970).

67. R. W. Baker and H. K. Lonsdale in F. W. Harris, Ed., *Proc. Int. Controlled Release Pesticide Symp., Ohio*, 1975, p. 9.

68. T. J. Roseman, in R. L. Goulding, Ed., *Proc. Int. Controlled Release Pesticide Symp. Oregon*, 1977, p. 403.

69. A. N. Neogi and G. G. Allan, "Controlled Release Pesticides: Concepts and Realization," *Advan. Exper. Med. Biol.*, **47**, 195 (1974).

70. B. L. Browning, *The Chemistry of Wood*, Interscience, New York, 1963, p. 137.

71. T. S. Gardner and C. B. Purves, *J. Am. Chem. Soc.*, **64**, 1539 (1942).

72. J. Honeyman, *J. Chem. Soc.*, **1947**, 168.

73. G. G. Allan, U.S. Patent 3,813,236 (1974).

74. H. Morawatz and P. E. Zimmering, *J. Phys. Chem.*, **58**, 753 (1954).

75. K. E. Savage, C. L. McCormic, and B. H. Hutchinson, in F. E. Brickman and J. A. Montemarano, Eds., *Proc. 5th Int. Symp. Controlled Release Bioactive Materials, Ohio*, 1978. p. 3.18.

76. C. L. McCormick, D. K. Lichatowich, and M. M. Fooladin, in F. B. Brickman and J. A. Montemarano, Eds., *Proc. 5th Int. Symp. Controlled Release Bioactive Materials, Ohio*, 1978, p. 3.6.

ADDITIONAL READING

1. N. Cardarelli, *Controlled Release Pesticides Formulations*, CRC Press, Cleveland, 1976.
2. J. R. Robinson, Ed., *Sustained and Controlled Release Drug Delivery Systems*, Marcel Dekker, New York, 1978.
3. R. Baker, Ed., *Controlled Release of Bioactive Materials*, Academic, New York, 1980.
4. R. L. Julians, Ed., *Drug Delivery Systems: Characteristics and Biomedical Applications*. Oxford Univ. Press, London, 1980.
5. J. C. Johnson, Ed., *Sustained Release Medications*, Noyes Data Corp., NJ, 1980.
6. A. F. Kydonieus, Ed., *Controlled Release Technologies: Methods, Theory and Applications*, Vols. I and II, CRC Press, Florida, 1980.
7. *Proceedings of the International Symposia on Controlled Release of Bioactive Materials* (1st to 8th, 1974–1981).

CHAPTER FOUR

Physical Methods of Controlled Release

L. T. ZEOLI
A. F. KYDONIEUS
Hercon Division
Health-Chem Corporation
New York, New York

CONTENTS

1. INTRODUCTION

In a controlled-release system, a biologically active material is emitted from a reservoir in a dispenser or other device at a predetermined rate for a specified period of time. In the face of a growing awareness during the last decade that substances ranging from drugs to agricultural chemicals are often excessively toxic, or sometimes ineffective, when administered by conventional means, workers have been turning to controlled-release technology to improve their products. When conventionally administered drugs in the form of pills, capsules, and injectables are introduced into the body, they enter as pulses and usually large fluctuations in concentration of the drug in the bloodstream and tissues and, consequently, unfavorable patterns of performance, that is, low efficacy and elevated toxicity. Applied conventionally, agricultural chemicals similarly provide initial concentrations far in excess of what is required for immediate results in order to assure the presence of sufficient chemical for a practical time period; such overdosing wastes much of the chemical's potential and all too often injures nontarget organisms.

The process of molecular diffusion through polymers and synthetic membranes has been found to be an effective and reliable means of attaining the controlled release of a variety of active agents. Central to the development of controlled delivery systems for drugs is the synthesis of the principles of molecular transport in polymeric materials and those of pharmacokinetics and pharmacodynamics. Pharmacokinetics is an important consideration because target tissues are seldom directly accessible, and drugs must be transported from the portal of entry into the body through a variety of biological interfaces before they reach the desired receptor site. During the transport process, the drug can be altered or degraded extensively and thereby produce a pattern of delivery at the receptor site markedly different from the pattern of drug release into the body.

Agricultural chemicals similarly suffer decreases in concentration and persistence at the desired sites following application as a result of biodegradation, chemical degradation, photolysis, volatilization, surface runoff, and leaching.[1]

1.1. Advantages and Limitations of Controlled Release

The advantages of controlled release can be great.[2-5] Controlled-release systems have already been used to achieve important benefits in dispensing active agents, thus elevating many products to commercial success.

Conventionally, fertilizers and insecticides and other pesticides are applied periodically to crops by broadcasting, spraying, and so on. Initial concentrations are very high and sometimes toxic to humans, precluding reentry of treated areas for extended periods of time. Subsequently, concentrations of chemicals diminish rapidly and soon fall below the minimum effective level. The problem is magnified by recent requirements that agricultural chemicals be biodegradable or at least nonpersistent in the environment. As a result, chemicals must be applied repeatedly to maintain control.

The principal advantages of controlled-release formulations are that they allow much less pesticide to be used more effectively for a given time interval, and they are specifically designed to counter the environmental processes that degrade conventionally applied pesticides. In most instances, the rate of degradation follows first-order kinetics.[6-9] Thus, if M_e is the minimum effective level, M_∞ the amount of agent applied initially; and k_r the rate constant, then t_e, the time during which an effective level of pesticide is present after a single application, would be

$$t_e = \frac{1}{k_r} \ln \frac{M_\infty}{M_e}$$ (1)

It follows from Eq. 1 that to increase the duration of effective action of a conventionally applied pesticide, the application of an exponentially greater quantity of the pesticide is required. If, however, the pesticide could be maintained at the minimum effective level by a continuous supply from a controlled-release system, then optimum performance of the insecticide would be realized, and the duration of action, t_e, would be

$$t_e = \frac{M_\infty - M_e}{k_d M_e}$$ (2)

where k_d is the rate constant for pesticide delivery from the controlled-release device.

Despite its advantages, controlled release is not always a panacea; negative effects may at times more than offset the benefits. Some of the disadvantages of controlled release, and other considerations that require thorough appraisal, are:

1. The cost of the controlled-release preparation and processing, which may be substantially higher than that of standard formulations.
2. The fate of the polymer matrix and its effect on the environment.
3. The fate of polymer additives such as plasticizers, stabilizers, antioxidants, and fillers.
4. The environmental impact of the polymer degradation products following heat, hydrolysis, oxidation, solar radiation, and biological degradation.
5. The cost, time, and probability of success in securing government registration of the product, if such is required.

Once an application has been thoroughly investigated and found suitable, one must select (1) the controlled-release technology that best suits the application and (2) the basic physical form of the device, which includes selection of the polymer matrix for controlling the release of the active agent.

1.2. Basic Components of Controlled-Release Devices

A controlled-release product consists of the active agent and the polymer matrix or matrices that regulate its release. Although pharmaceutical, agricultural, and veterinary chemicals have been used as active agents in commerical and experimental controlled-release devices,[10,11] the major effort in this field has been directed toward the administration of pharmaceuticals and the application of pest-control chemicals.[11]

Advances in controlled-release technology in recent years have been rapid because polymer science has become sophisticated enough to incorporate into polymers tailor-made properties for each controlled-release application. In selecting the polymer matrix, the following design criteria should be considered:

1. Molecular weight, glass transition temperature, and chemical functionality of the polymer must allow the proper diffusion and release of the specific active agent.

2. Polymer functional groups should not react chemically with the active agent.

3. The polymer and its degradation products must be nontoxic to the environment, and in medical applications neither toxic nor antagonistic to the host.

4. The polymer must not decompose in storage nor during the useful life of the device.

5. The polymer must be easily manufactured or fabricated into the desired product, and should allow incorporation of large amounts of active agents in the product without deteriorating its mechanical properties.

6. The cost of the polymer should not be excessive, which would cause the controlled-release device to be noncompetitive.

A list of polymers that have been used in controlled-release formulations is shown in Table 1.[4,10-14] These polymers are used in coatings, in microencapsulation, as films in laminated structures, as slabs in monolithic systems, and as flakes in many erodible devices. Figure 1 illustrates some of the forms that controlled-release products can take.

1.3. Basic Release Characteristics of Controlled-Release Processes

All controlled-release processes (Table 2) are in one way or another controlled by the diffusion of the active agent through a polymer barrier or by an inward

TABLE 1. Polymers Used in Controlled-Release Devices

Natural Polymers	
Carboxymethylcellulose	Zein
Cellulose acetate phthalate	Nitrocellulose
Ethylcellulose	Propylhydroxycellulose
Gelatin	Shellac
Gum arabic	Succinylated gelatin
Starch	Waxes, paraffin
Bark	Proteins
Methylcellulose	Kraft lignin
Arabinogalactan	Natural rubber

Synthetic Polymers	
Polyvinyl alcohol	Polyvinylidene chloride
Polyethylene	Polyvinyl chloride
Polypropylene	Polyacrylate
Polystyrene	Polyacrylonitrile
Polyacrylamide	Chlorinated polyethylene
Polyether	Acetal copolymer
Polyester	Polyurethane
Polyamide	Polyvinylpyrrolidone
Polyurea	Poly(p-xylylene)
Epoxy	Polymethyl methacrylate
Ethylene-vinyl acetate copolymer	Polyhydroxyethyl methacrylate
Polyvinyl acetate	

Synthetic Elastomers	
Polybutadiene	Acrylonitrile
Polyisoprene	Nitrile
Neoprene	Butyl rubber
Chloroprene	Polysiloxane
Styrene-butadiene rubber	Hydrin rubber
Silicone rubber	Ethylene-propylene-diene terpolymer

diffusion of an environmental fluid in the case of "environmental agent ingression" devices and some homogenous "retrograde chemical reaction" devices.

Possibly the only systems that are not controlled by diffusion are the "pure" erodible devices and some heterogeneous retrograde chemical reaction devices, but even in these systems release is generally achieved through a combination of diffusion and erosion.

The active agent passes through the polymeric barrier in the absence of pores or holes by a process of absorption, solution, and diffusion down a gradient of thermodynamic activity until desorbed or removed. The transport of the active agent is governed by Fick's first law:

$$J = \frac{dM_t}{Ad_t} = \frac{-D\,dC_m}{dx} \qquad (3)$$

where J is the flux in g/cm^2 · s, A is the surface area through which diffusion takes place in cm^2, M_t is the mass of agent released in g, dM_t/dt is the steady-state release rate at time t, C is the concentration of active agent in the polymeric membrane in g/cm^3, dC_m/dx is the concentration gradient, and D is the diffusion coefficient of the active agent in the polymeric membrane in cm^2/s.

Equation 3 can be integrated under the proper boundary conditions for each of the systems listed in Table 2 to obtain a formula for the amount of agent released as a function of time. In many situations, however, the mathematics becomes rather complicated, and no explicit equations can be derived. For more information on the mathematics of diffusion, see references 15-17.

2. DESCRIPTION OF SPECIFIC SYSTEMS

Table 2 categorizes the various controlled-release systems broadly as either *physical* or *chemical*. A description of the individual physical controlled-release technologies follows.

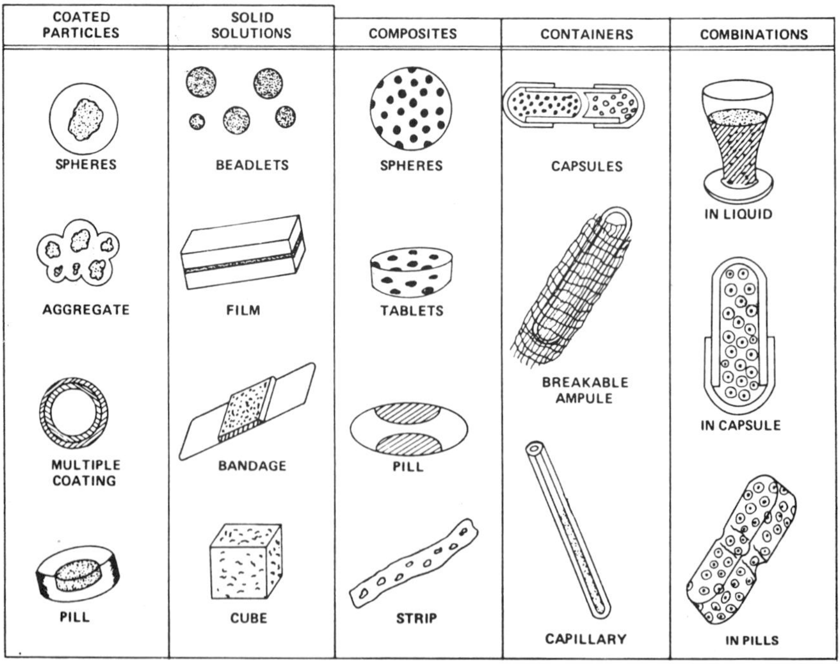

FIGURE 1.   Various types of controlled-release devices.

TABLE 2. Categorization of Controlled-Release Polymeric Systems

I. Physical Systems
  A. Reservoir systems with rate-controlling membrane
    1. Microencapsulation
    2. Macroencapsulation
  B. Reservoir systems without rate-controlling membrane
    1. Hollow fibers
    2. Poroplastic® and Sustrelle™ ultramicroporous cellulose triacetate
    3. Porous polymeric substrates and foams
  C. Monolithic systems
    1. Physically dissolved in nonporous polymeric or elastomeric matrix
      a. Nonerodible
      b. Erodible
      c. Environmental agent ingression
    2. Physically dispersed in nonporous polymeric or elastomeric matrix
      a. Nonerodible
      b. Erodible
      c. Environmental agent ingression
  D. Laminated structures
    1. Reservoir layer chemically similar to outer control layers
    2. Reservoir layer chemically dissimilar to outer control layers
  E. Other Physical Methods
    1. Osmotic pumps
    2. Adsorption onto ion-exchange resins
II. Chemical Systems
  A. Chemical erosion of polymer matrix
    1. Heterogeneous
    2. Homogeneous
  B. Biological erosion of polymer matrix
    1. Heterogeneous
    2. Homogeneous

2.1. Reservoir Systems with Rate-Controlling Membrane

These include micro- and macrocapsules. Microencapsulation is a procedure that reproducibly applies a uniformly thin polymeric coating (rate-controlling membrane) around small solid particles, droplets of liquid, or dispersions of solids in liquids, with the size of the resulting capsules ranging from a few tenths of a micrometer to several thousand micrometers.[2] Capsules greater than 2000–3000 μm are called macrocapsules. There are no real differences between micro- and macrocapsules with respect to the release characteristics or type of active agent that can be encapsulated. Many methods for microencapsulating active agents have been developed; these techniques are discussed elsewhere in this book.

2.1.1. Films and Membranes

Another controlled-release device that falls into this category is the polymeric film membrane. These devices function by the diffusion of active molecules from a reservoir to the environment. The membrane provides a predetermined resistance to outflow of chemicals that is a function of film thickness, film composition, the migrating species, and the given environment.

Polymeric films have been developed for such specialized applications as protective packaging, reverse osmosis, dialysis, hyperfiltration, and ion exchange. In some highly specialized controlled drug-delivery systems, the rate-controlling membrane may interface with the biological site and often is in intimate contact with delicate tissues.

Release characteristics. Fick's law predicts that a steady state will be established, with the release rate being constant and independent of time, if an active agent is enclosed within an inert polymer membrane and thermodynamic activity of the agent is maintained constant within the enclosure. The amount of active agent released per day is therefore constant for the life of the device, and

$$M_t = kt \qquad (4)$$

where k is a constant. Equation 4 applies for all geometries of the device—spheres, slabs, and so on. This zero-order type of release is illustrated by curve I in Fig. 2.

The development of controlled-release technology motivated a series of studies on polydimethylsiloxane (PDMS) tubing, in which anesthetics, thyroid hormone, and certain cardiovascular agents were released in animals.[18-22] The promising results led to work in related areas of pharmacology.

The principal developments in the use of silicone-rubber membranes have involved the controlled release of steroids, primarily for contraception. PDMS capsules implanted subdermally in women could, if successful, provide 5-year contraceptive activity.[23-27] In studies directed toward the development of a contraceptive system for men, PDMS capsules that release testosterone have been implanted in male rabbits to demonstrate that sperm production can be successfully countered for months.[28,29]

Careful study of the performance characteristics of the many polymeric materials that are available have increasingly focused attention on the ethylene vinyl acetate (EVA) polymers. For certain applications, these thermoplastic materials were found superior to silicone rubber. For example, because certain members of the EVA polymer family are flexible without plasticizers, EVA copolymers are candidates for replacing the plasticized polyvinyl chloride used in

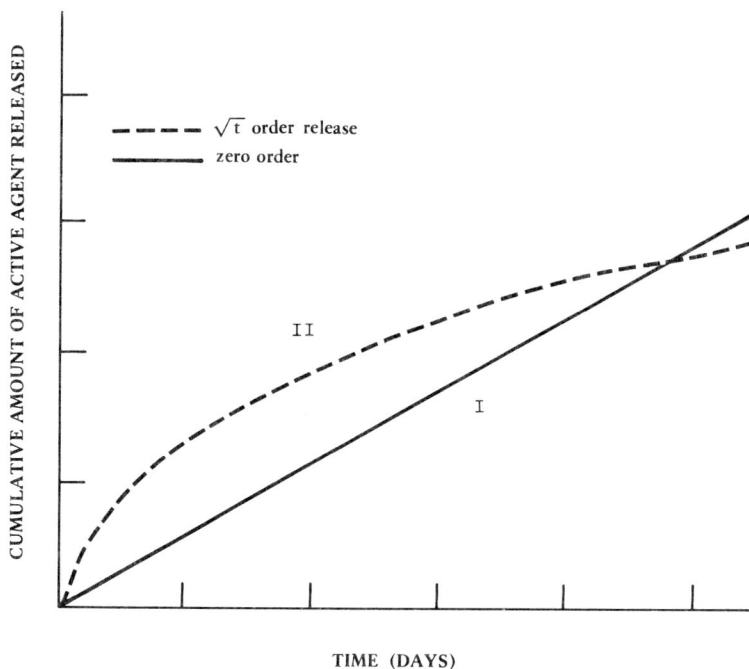

TIME (DAYS)
FIGURE 2. Zero-order and $\sqrt{t}$-order release.

blood containers and tubing. Other advantages of the EVA polymers are that they are chemically stable and can be easily sterilized.

Clinical studies conducted by the Alza Corporation have lead to the commercial introduction of an ethylene vinyl acetate membrane-controlled intrauterine progesterone contraceptive system.[30] Another membrane-controlled therapeutic system that utilized an EVA copolymer slowly releases medication for the treatment of the elevated intraocular pressure associated with chronic glaucoma.[31]

Polyurethanes have also been studied for biomedical applications,[32] for example, for use in artificial kidney elements that separate molecules on the basis of chemical structure as well as physical size. Other hydrophobic polymers found useful as rate-moderating membranes are polyethylene,[33-35] polyamide,[36-39] polymethyl methacrylate, and poly-n-butyl methacrylate.[40]

2.2. Reservoir Systems without Rate-Controlling Membrane

These systems include hollow fibers,[41-43] chemical-impregnated porous plastics such as MPS porous PVC sheet, millipore fibers, and Cellgard porous polypropylene,[44,45] Foams and possibly hydrogels,[46-50] and ultramicroporous cellulose triacetate.[51,52]

2.2.1. Hollow Fibers

The simplest example is perhaps the hollow fibers that hold the active agent in tiny open tubes from which the agent escapes to the outside by diffusion through through the air layer above it. Systems utilizing impregnated porous plastics (PVC, Cellgard, etc.) are more complex, but in all cases the active agent is retained by capillary action or physically embedded in the pores. Release also occurs by diffusion to the outside through the air layer above the liquid in the pores. Stricly speaking, most of these systems may be considered monolithic matrix systems except that interaction of active agent and polymer is minimal.

Release characteristics. If a material is allowed to evaporate from the lumen of a hollow fiber sealed at one end and open at the other, the release curve of mass released vs. time is characterized by a steep slope initially, then by an extended lower flat region which approximates zero-order release kinetics. An example of this kind of release is given as curve II in Fig. 2.

The amount of agent released is proportional to $\sqrt{t}$, as expressed by the equation

$$M_t = k\sqrt{t} \tag{5}$$

Release rates of hollow fibers can be manipulated by varying the internal diameter of the fiber or, for a cluster of fibers, by varying the number of fibers in the cluster. The longevity of the dispensers is also directly proportional to the length of the column of active material in the lumen reservoir at time zero and may depend upon the composition of the fiber itself. The mechanism for the release of a volatile material comprises three steps:

1. Evaporation of the liquid–vapor interface.
2. Diffusion from the liquid–vapor interface to the open end of the lumen.
3. Movement to the outside away from the open end.

In most cases the diffusion step is the rate-controlling one.[53]

Like every other controlled-release system, hollow fibers have distinct technical and economic limitations. Hollow fibers are limited by (1) an intrinsically low mass capability, (2) a high ratio of fiber weight to active material weight, and (3) the requirement for specialized application equipment. To date, hollow fiber formulation studies have focused almost exclusively on insect phermones as alternatives to conventional pesticides. Such efforts have shown the enormous potential for properly formulated phermones in pest management programs designed to protect crops or forest land while minimizing the stress of toxic chemicals on the environment.

2.2.2. Proplastic® and Sustrelle™ Ultramicroporous Cellulose Triacetate

Another controlled-release system without a rate-controlling membrane that is beginning to receive broad commercial acceptance is an ultramicroporous open-celled form of cellulose triacetate.[54-56] The film or membrane configuration has the trade name Poroplastic®; the powder or microbead form is trade named Sustrelle™. The material can also be used to coat other substances or to make hollow fibers. Experiments on the film and powder have shown that the difference in release rates between the two forms can be explained by the difference in surface area-to-volume ratio between the film and powder.

These products are essentially molecular sponges with a cellulose triacetate matrix. Because the pore dimensions are extremely small, the internal surface area is very large, and large quantities of liquid can be strongly retained within the material by capillary action.

The properties of Poroplastic and Sustrelle, some of which are unique, are listed below:[57]

1. Extremely small yet variable pore size. Characteristic pore diameters range from 14 Å to over 250 Å.

2. High liquid content. Liquid content can be adjusted from 70 to 98%; this holds true for almost any kind of solution.

3. Irreversible shrinkage on drying. As the liquid in Poroplastic evaporates, the pore structure progressively collapses. The process is irreversible because the swollen state is not thermodynamically preferred over the partially or completely dry state.

4. Homogeneous and transparent. The hydrolytic permeability and the diffusive permeability of Poroplastic are inversely proportional to thickness. Also, because of the extremely small size of the pores and the lattice structure, Poroplastic does not scatter light, but is transparent.

5. Stong, crystalline, and noncross-linked.

The release characteristics of Poroplastic are consistent with the model of an open-cell molecular sponge made of cylindrical micellular units. In accordance with the model, flux is directly proportional to applied pressure and inversely proportional to thickness.

Sustained release from ultramicroporous cellulose triacetate is basically diffusion-controlled. The CR process can be considered to occur in three steps: (1) impregnation, (2) fixation, and (3) diffusional release.

The impregnation step involves a series of diffusional exchanges of miscible liquids until the liquid within the polymer matrix is the active agent to be released. The fixation step can be one of several procedures to slow the rate of release of the active agent, for example, pore collapse and precipitation of the active agent within the porous matrix. This precipitation can be achieved by solvent evaporation in a saturated solution, solvent substitution, or chemical reaction. Diffusional release often requires a trigger mechanism for initiation.

Many commercial applications have been developed for Poroplastic and Sustrelle. The types of applictions fall into one of two broad categories, those based on its characteristics as a matrix and those related to a controlled release.

As a matrix, the most important properties are those relating to its high diffusion coefficient and to its ability to hold large quantities of liquid within a solid form. Thus, the matrix has been used as a carrier for electrical insulating oils, radiation absorbers in solar energy collectors, flavor and fragrance modifiers, photochromic solutions for eye goggles, indicators for toxic substance monitoring, and solvent extraction systems in hydrometallurgy.

The controlled-release applications usually depend on the product's regulated pore size, shrinkage, deformation, and liquid or vapor expulsion. It has been used for the controlled release of pharmaceuticals, pesticides, air freshners, dermatological products, toiletries, surfactants, shelf life extenders, cosmetics and perfumes, and adhesives, and in analytical procedures such as chromatography, ultrafiltration, enzyme or antibody immobilization, and electrophoresis.

2.2.3. Hydrogels

Some hydrophilic monomers form polymeric materials that absorb substantial amounts of water, forming three-dimensional networks with an equilibrium water content; such materials are known as hydrogels. Although natural hydrogels have received little attention as rate-controlling membranes, some synthetic polymers, especially the hydroxyalkyl acrylates, offer great promise for such use because of the ease of adjusting their permeability via the degree of hydration, a property that can be varied by changing the comonomer ratios, cross-linking agent concentrations, and conditions of polymerization. Transport accross a hydrogel membrane is largely a function of the water solubility of the agent and involves primarily the entrapped aqueous phase rather than dissolution of the agent in the polymer itself. Further information on synthetic hydrogels can be found in a review article by Ratner and Hoffman.[58] Studies establishing the biocompatibility of materials such as 2-hydroxyethyl methacrylate have resulted in their utilization in soft contact eye lenses, as catheter coatings, and as burn dressings.

The ease of synthesis of copolymers from hydrophilic acrylic monomers has simplified the preparation and study of such membranes. The membranes have a wide range of properties and are potentially suitable as rate-controlling membranes for water-soluble agents. Theoretical relationships for the permeability of hydrogels to water and solutes in terms of the free volume in the membrane have been advanced by Yasuda, Lamaze, and Peterline.[59] Chen[60] observed that the concentration of cross-linking agent has an effect upon water uptake. The behavior of water solutes in hydroxyalkyl methacrylate polymers with low cross-link density can be explained in terms of a very loose network that is porous on a submicroscopic level; pore radii ranging from 3 to 50 Å have been calculated.[61] However, because the diffusion coefficient of water approached a limiting value as the concentration of cross-linking agent exceeded approximately 5 mol %, transition to a partition-type mechanism was proposed to explain the phenomenon.[62]

The most extensive work to date in which hydrogels are employed as rate-moderating membranes has been associated with the development of controlled-release systems for dispensing fluoride salts in the mouth over long periods of time for the prevention of dental caries. In the most successful approach, the fluoride was dispersed in a solid hydrogel reservoir surrounded by a coating of another hydrogel through which the fluoride permeated more slowly.[63] This work, which is part of a major effort by the National Institute of Dental Health, is still at the stage of animal studies.

Several researchers at Hydromed (a subsidiary of National Patent Development) have reported the preparation of three types of hydrogel-membrane-coated systems for releasing the narcotic antagonist cyclazocine. a drug useful in the rehabilitation of narcotics addicts. Some of the systems provided controlled release of the drug for longer than one month, compared to about 6 days for the uncoated drug.

The diffusion of several anticancer drugs through hydrogels[64] has been studied to aid in the development of pouch-shaped devices for placement directly on tumors. Other groups have incorporated antibiotics into hydrogels for sustained release.[65-68] Hydrogel contact lenses have been impregnated with drug solutions (pilocarpine, epinephrine, etc.) for prolonged release to the eye.[69,70] However, the principle of interposing a barrier membrane with hydrogels has not been investigated for such applications.

An investigation in which the surface of the hydrogel is modified is currently under way at the University of California.[71] It involves *CASING* (cross-linking by active species of inert gases), whereby the surface of the hydrogel is modified to create a barrier membrane. The problem with these laminates is that the surface film is apt to fracture as the water-swollen hydrogels swell and shrink, making the practicality of this approach doubtful.

Hydrogel coatings have also been applied to other polymeric substrates. Scott et al. observed that these coatings could prolong the activity of copper-bearing intrauterine devices as well as improve their biocompatibility.[72]

Natural hydrogels include polysaccharides such as cross-linked dextrans

(Sephadix®) and cellulose (Cuprophan®), and polypeptides such as collagen. Cuprophan has become the most widely used material for hemodialysis membranes, and considerable literature has appeared on its permeability to molecules of biomedical interest.[73-77] Collagen, the chief organic constituent of connective tissues and the most common polymer in the body, is the subject of increasing attention as a film-forming polymer and has been used for surgical sutures as well as in nonmedical applications.[78] A collagen membrane per se has not yet been reported.

2.2.4. Impregnated Filters

A number of films that provide discrete, continuous, very fine pores ranging from several micrometers to a few angstroms in size have recently become available (see Table 3). All are highly permeable to gases and water vapor; however, the barrier properties of each film depend largely upon the agent with which the pores are filled. To date, microporous films have found the greatest use as filters and as separators for battery compartments.[79] Relatively little has been published on microporous membranes as rate-controlling membranes between a reservoir and a sink.

The single major application of a microporous rate-controlling membrane has been in the transdermal therapeutic system (TTS) developed by the Alza Corporation. Essentially, it is a device designed to deliver a drug through the skin at a controlled rate.[80] The TTS is a multilayer laminate consisting of a drug reservoir containing the active agent dispersed as a separate phase within a highly drug-permeable matrix, laminated between the rate-controlling membrane (which rests on the skin) and an external metallic foil, which is impermeable to the drug and moisture. The resulting pattern of drug in the body corresponds to that achieved by infusion.

TABLE 3. Some Commercial Microporous Films[81]

Base Polymer	Range of Pore Size (Å)	Trade	Source
Regenerated cellulose	1000	Millipore	Millipore
Cellulose nitrate/acetate	1000	Diapor	Amicon
Cellulose triacetate	15–250	Poroplastic	Moleculon
Polypropylene	100–2000	Celgard	Celanese
Polytetrafluoroethylene	1000	Gore-Tex	W. L. Gore Associates
Polycarbonate	300	Nucleopore	General Electric

2.3. Monolithic Systems

Probably the simplest and least expensive means of controlling the release of an active agent is its uniform dispersal in an inert polymeric matrix. Generally, the active agent is physically blended with the polymer powder and the mixture is then fused together by compression molding, injection molding, screw extrusion, calendering, or casting,[82] all of which are common processes in the plastics industry. Alternatively, the active agent is blended with elastomeric materials in the mixing step as is done with the other additives, such as accelerators, reinforcing pigments, stabilizers, and processing aids.[83]

In either case, the active agent dissolves in the polymeric or elastomeric matrix, generally until saturation is reached. Additional active agent, if any, remains dispersed within the polymer matrix after it is physically dispersed and molded or cured. As the agent evaporates or is otherwise removed from the surface of the monolithic device, more agent diffuses out from the interior to the surface of the monolithic device, more agent diffuses out from the interior to the surface in response to the decreased concentration gradient leading to the surface.

If the polymer used is soluble in ambient fluids or degrades during its intended use, the monolithic device is erodible. In this case, the active agent is released by a combination of diffusion and liberation due to erosion.

Pure degradable and erodible systems, which can be defined as those that release their contents by diffusion, osmotic bursting, leaching, and any other controlled-release mechanism, can have the very desirable property of chemically or biologically degrading after the useful life of the device has expired. Contamination of the environment with such devices is apt to be minimal.

Erodible systems can be considered a subdivision of degradable systems. They release their contents by physically "uncovering" the active agent as chemical or biological erosion of the matrix occurs. Erodible and degradable systems have been discussed more fully in various publications.[14,84-86]

If the polymer can be plasticized or made to swell by an environmental agent such as water, the environmental agent invades the device, plasticizes the polymeric matrix, and thereby allows the physically bound active agent to diffuse out. Such systems, which include starch xanthate[87,88] and perhaps some lignin-modified devices,[89] differ from erodible ones in that the matrix remains physically intact.

2.3.1. Physically Dissolved in Nonporous Polymeric or Elastomeric Matrix

A. Physically Dissolved, Nonerodible Polymeric or Elastomeric Matrix. The release rate in such systems is proportional to $t^{-\frac{1}{2}}$ ($\sqrt{t}$ order) until about 60% of the active agent is released. Thereafter it is related exponentially to time:

$$\frac{dM_t}{dt} = k_1 e^{-k_2 t} \tag{4}$$

where k_1 and k_2 are constants. Thus, beyond the 60% level, the rate of release drops exponentially. This type of release (first-order release) is also observed in reservoir systems in which the solution of active agent within the enclosure is less than saturated.

B. Monolithic Erodible Sytems. If one assumes that release of the active agent by diffusion is negligible in this system, the speed of erosion will control the release rate. Release by erosion is a surface area-dependent phenomenon, and the release will be constant (zero-order) as long as the surface area does not change during the erosion process. This is essentially true for slab-shaped devices. Eroding cylinders and spheres give delivery rates that decrease with time (owing to decreasing surface area), even though the kinetic process providing the rate-determining step is, in fact, zero-order.[90]

One obvious advantage of the erodible (biodegradable) devices is that after an implanted device has fulfilled its function the matrix does not require surgical removal. For a polymer to be used as the matrix in such a device, it must (1) be compatible with the environment and with the active agent, (2) have a rate of biodegradability compatible with the required life span of the device, (3) be able to provide the desired release rate, and (4) be readily available.

Some of the methods that have been used to prepare these devices include (1) dissolving the polymer and the active agent with or without plasticizer in a solvent, evaporating the solvent, and press-melting the residue plates to produce a film;[91,92] (2) grinding the polymer, mixing it with the active agent, and compression molding or extruding the mixture into films or pellets;[93] and (3) covalently bonding the drug with reactive groups of the polymer and preparing films as above.[94] Films have also been reformed into chips, beads, tubes, and rods.[93,95]

Polyactic acid (PLA) has been found to be a suitable material for surgical implants because it undergoes hydrolytic deesterification to lactic acid, a normal product of muscle metabolism.[91,95-100] PLA has also been used for preparing absorbable sutures.[101]

PLA has been most widely used with narcotic agents,[93,102,103] fertility control agents,[95,104] anticancer agents,[100] and herbicides and pesticides.[105]

Polyglycolic acid (PGA) has also been used in the preparation of sutures (Dexon®),[98,106,107] prosthetic devices,[108] and storage pellets.[109] However, its use is limited because of difficulties in fabricating composites, for example, its low solubility in common solvents.

The copolymer PLA/PGA has also proved to be a useful matrix in the preparation of controlled-release devices for drugs, fertilizers, and insecticides.[110-117]

Parameters influencing the release characteristics of monolithic devices can be classified as solute-dependent or solute-independent. Solute-dependent parameters are related to the physicochemical nature of the solute in the polymer, while the solute-independent parameters are system variables. Either can be varied to optimize the delivery rate of solute, but in pharmaceuticals, for example, the chemical composition of the polymer is usually varied to achieve the desired release rate. The various factors that affect the release rate characteristics of monolithic devices are summarized in Table 4.

The effect of *concentration* on the release of steroids from silicone rubber has been amply investigated.[118-122] For monolithic devices that contain the dispersed drug, release rates are proportional to the square root of the concentration. Release of the drug from monoliths containing dissolved drug is expected to be linearly related to the initial concentration. This concept has been less extensively studied, but the linear dependence predicted has not always been noted.[123-130] This suggests that another variable may have been altered when the concentration was changed.[131]

The *diffusion coefficient* is a function of the molecular size and shape of the diffusing molecule, the degree of polymer crystallinity, chain–chain interactions, and flexibility.[132-134] Only limited data are available on diffusion coefficients of drugs in polymers, but the wide range of values indicates that the diffusion coefficient is a major factor in determining the rate of solute delivery. Release rate is found to be proportional to the square root of the diffusion coefficient.

In the case of membrane transport, the product of the *partition coefficient* and the diffusion coefficient defines the permeability constant. The partition coefficient is an additive property of the functional groups present in a molecule[135] and is extremely sensitive to slight changes in molecular structure. Roseman and Yalkowsky[136] demonstrated that the overall release kinetics are biphasic; the early portion is independent of time (zero-order), and the later portion is

TABLE 4. Factors Affecting the Release Characteristics of Monolithic Devices

Factors Dependent upon Solute
  Solubility
  Partition coefficient
  Diffusion coefficient

Factors Independent of Solute
  Concentration
  Geometry
  Tortuosity of pores
  Porosity
  Volume fraction
  Diffusion layer

dependent on the square root of time. The magnitude of the partition coefficient is a controlling factor in determining the duration of the zero-order period.

Solubility of the solute in the polymer or elution media depends on the intermolecular forces between the solute and the solvent. Release of drugs from monolithic devices containing excess drug is proportional to the square root of their solubilities in the polymer. Chien[137] presented the thermodynamic relationships for the dependency of the solute's solubility in the polymer on its melting point. This was expressed by the following equation:

$$J_{max}h\exp(1+\chi) = \frac{-\Delta S_f}{R}\left(\frac{T_m}{T}-1\right) + \ln\rho_d D_s \qquad (5)$$

where J_{max} the flux from a saturated solution across a membrane of thickness h, χ the solute–polymer interaction constant, ΔS_f the entropy of fusion, R the gas constant, T_m the melting point of the polymer, ρ_d the density of the solute, and D_s the distribution coefficient of the polymer. The interaction constant depends on the solubility parameter and is given by

$$\chi = \frac{V_d}{RT}(\delta_d - \delta_p)^2 \qquad (6)$$

where δ is the solubility parameter and the subscripts d and p refer to the solute and polymer, respectively.

The overall model for drug release from a dispersion monolith is a four-step process:

1. Dissolution of drug particles within the polymer.
2. Diffusion of dissolved drug through the polymer phase.
3. Partitioning into the eluting solvent, with subsequent diffusion across the boundary diffusion layer.
4. Absorption acorss the biological membrane.

Porous monoliths were developed to be ingested orally (i.e., tablets), while homogeneous monoliths have been used to dispense a variety of drugs into animal or human body cavities. Upon contact with body fluids, the release process is activated, and the drug diffuses through the polymer or channels within its structure to the surface, where absorption takes place.

Synthetic polymers can be classified as elastomers (rubbers) or plastomers (plastics). Elastomers possess relatively weak interchain forces, giving rise to their elastic properties. Plastics, on the other hand, have moderate interchain forces, so that if such molecules are extended by means of an applied outside force, that orientation is retained after the external force is removed. An intermediate group of materials with both elastic and plastic properties does exist and was once referred to as "elastoplastics." Ethylene-propylene-diene terpolymers, polyethylene vinyl acetate, polyvinyl acetate, certain modified acrylics, and others fall in this class.

2.3.2. Physically Dispersed in Nonporous Polymeric or Elastomeric Matrix

The release rate in this system is proportional to $\sqrt{t}$ as long as the concentration of the active agent present (dispersed and dissolved) is higher than the solubility of the agent in the matrix. Thus, release from the dispersed system is similar to that from the dissolved systems, except that the release rate does not decrease after 60% of the chemical has been released; instead, the relationship holds for almost the complete release curve.

A. Nonerodible. The first known product based on the monolithic dispersal of an insecticide in polyvinyl chloride (PVC) was the Shell No-Pest® strip. The active agent, dichlorvos (dimethyl 2,2-dichlorovinyl phosphate), is released through diffusion, its emitted vapors being insecticidal.[79] Emission of the active agent varies with the type and amount of plasticizer. Consequently, the presence of a plasticizer may increase or decrease the loss of the active agent. Processing problems may be encountered if the material is highly volatile, and materials of low volatility may not be efficient for a given use because the rate of emission may be too low. To date, only a few products consisting of a pesticide monolithically dispersed in a plastic dispenser have reached the commercial stage.

B. Erodible. Recent developments have made possible the incorporation of plant and animal trace nutrients into controlled-release devices. Elements essential to nutrition in most higher plants and animals (classified as trace nutrients) are sulfur, calcium, boron, copper, iron, manganese, molybdenum, zinc, cobalt, magnesium, selenium, silicon, fluorine, and iodine.

The problem associated with the conventional application of the nutrient to the target organism are similar to those found in the use of pesticides. Conventional treatment is generally wasteful of materials, higher in cost, and lower in efficiency than treatment using controlled release of the active materials.

Environmental Chemicals has incorporated nine elements into monolithic plastic matrices. The purpose of this effort was to formulate one or more chemicals in a plastic matrix so they would be released in moist environment (soil) at a rate that would satisfy (plant) nutritional requirements efficiently. The successful release of inorganic ions from a plastic matrix hinges on either the use of a porosity-enhancing coleachant or "porosigen" or the use of a mix of plastic materials. Aside from the development of materials of commercial interest, this work has provided insights into means of devising leaching-type systems.

Controlled-release monolithic elastomeric devices may operate via a diffusion–dissolution or a leaching–release mechanism. Three factors govern the type of system desired: (1) the solubility limit of the agent in the matrix, (2) dispenser geometry, and (3) desired use.

Since the establishment of a diffusion–dissolution mechanism depends on solubility of the active agent in the polymer matrix, the choice of base polymers is largely limited to those with adequate solvation power for the specific agent. The degree of solubility is also affected by the chemical and physical structure of the polymer. In general, agents are less soluble in materials of high molecular weight and/or high cross-link density than in materials of the same general monomeric structure with low molecular weight and/or lower cross-link density.

In summary, the effective release time of a given diffusion–dissolution formulation is dependent on the (1) polymer, (2) type and concentration of regulant, (3) vulcanization conditions, (4) product geometry, and (5) agent loading.

Antifouling rubber, now available from B.F. Goodrich Company, Akron, Ohio, was the first controlled-release elastomer. This material is a solid-sheet formulation that releases its antifoulant by a diffusion–dissolution mechanism. Known as Nofoul®, it is used in the rubber covering of sonar domes and has been found to remain effective for as long as 9 years.

Typical antifouling paints sometimes use an elastomer base, and most commonly rely on a leaching-type system. As one would expect, the effective life of these leaching-type systems is much shorter than that of the elastomers.

Both diffusion–dissolution and leaching-type controlled-release formulations have been developed for use as molluscides. The development of long-term continuous controlled-release organotin molluscicides was initiated by Cardarelli in 1966.[138-140] This work resulted in the development of the BioMet SRM™ material and CBL-9B. Natural rubber is the preferred elastomer from the standpoint of cost and biodegradability. Several other metallic and non-metallic organic chemicals used as fungicides have been found to be soluble in natural rubber and other elastomeric materials, spurring continued research in this area.

A "pesticidal bait" requires three elements: a toxic agent, an attractant, and a binding material. In addition, chemotaxis must exist for the target animal. A bait has two agents: the toxicant, which is usually bound to prevent its release, and the attractant, which is emitted through dissolution, exfoliation, leaching, vaporization, or another process. Although such baits are not usually considered controlled-release materials, it is evident that they incorporate similar features. In both, a biological agent must be released slowly from the binding element.

The development and evaluation of controlled-release herbicides for aquatic weeds was a natural outgrowth of the antifouling and molluscicidal work. In initial work with the butyoxyethanol ester and the diemthylamine salt of 2,4-dichlorophenoxyacetic acid in monolithic elastomerid dispensers, release of the compound only in the target habitat rather than in the entire body of water was sought. Because elastomers can be produced in a variety of forms, for example, with density adjusted by means of inexpensive fillers, elastomeric strips were developed that "hovered" at the phytozone level being attacked. The initial small-scale field tests were successful[141] and have served as the motivation for additional work with other agents as well.

Comparatively little work has been done in the application to soils of controlled-release herbicides as a means of controlling weeds of agricultural significance.[141a] The rate of release of the toxicant from the binding matrix depends on the amount of moisture present, as occurs with conventional herbicides. The pH and adsorption properties of the soil are also important.[142,143] It has been observed that the presumptive Ct (concentration times time) relationship does not hold when aquatic weeds are exposed to ultralow concentrations of controlled-release herbicides. This phenomenon, known as the chronicity effect, has led to the hypothesis that in these cases the target organism is unable to detect very minute quantities of the toxicant and that therefore the normal protective mechanisms triggered by the presence of larger concentrations of the toxicant do not occur.[144,145]

2.4. Laminated Structures

A specialized form of the monolithic device has been developed, patented, and made commercially available by the Hercon Division, Health-Chem Corporation, New York, NY.

The Hercon® controlled-release system of dispensing active ingredients consists of several layers of laminated polymeric materials. In essence, the Hercon dispenser has the active ingredient impregnated in a layer between two outer plastic layers. A schematic illustration of a typical Hercon delivery system is shown in Fig. 3. The specially formulated inner layer serves as a reservoir for the active ingredient while protecting it from oxidation and degradation by the environment. The active ingredient can be any of a wide variety of biologically active compounds, such as insecticides, insect pheromones or sex attractants, fragrance oils, room air fresheners, and antibacterial agents. Some of these multilayered dispensers have now been marketed for several years.

Basically, the active agent migrates continually, due to an imbalance of chemical potential, from the reservoir layer through one or more initially inert outer layers to the exposed surface, which is thereby rendered biologically or physiochemically active. At the surface, the active agent is removed by volatilization, thermal or ultraviolet degradation, alkaline or acid hydrolysis, or mechanical contact by humans, insects, rainfall, wind, or other agents.

Construction and composition of the dispenser varies with the active agent and with the release rate and duration of effectiveness desired. One or both

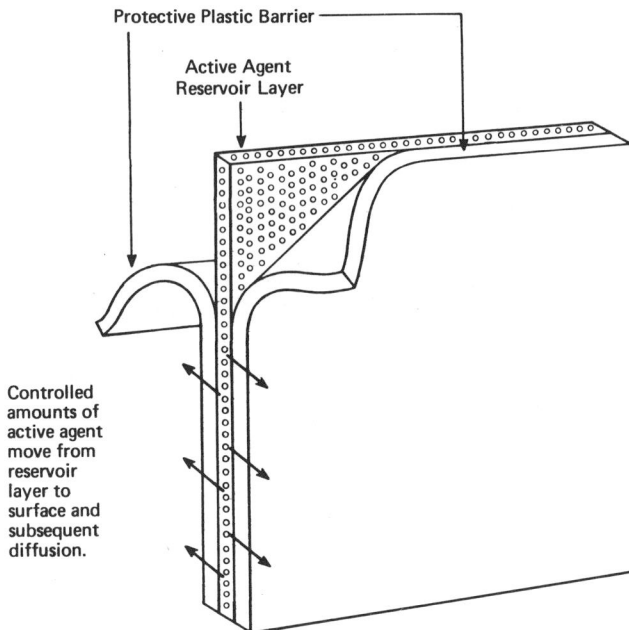

FIGURE 3.   Schematic diagram of a typical Hercon laminated dispenser.

outer surfaces of the dispenser may be made active by having one or both layers permeable to the active ingredient. The form of the dispenser may also be varied to aid in dispensing the active material. Thus, the original sheets may be cut into strips, ribbons, wafers, flakes, confetti, or even sprayable granules or powders.

Release characteristics. The concentration of the stored chemical and the composition and/or construction of the plastic layer components control the release rate of the chemical. Nonporous, homogeneous polymeric films, usually referred to as solution–diffusion membranes, are used in Hercon dispensers. Typical examples are silicone rubber, polyethylene, polyvinyl chloride (PVC), and nylon films.

The diffusant is able to pass through the membrane material in the absence of pores or holes by absorption, solution and diffusion down a gradient of thermodynamic activity, and desorption. The permeation process is governed primarily by Henry's law and Fick's first law.[146,147] Fick's first law states that the rate of transfer of diffusing substance through unit area of a section is proportional to the concentration gradient measured normal to the section:

$$J = -D\frac{dC_m}{dx} \tag{7}$$

where J is the flux in $g/cm^2 \cdot s$, C_m the concentration of diffusant or permeant in the membrane in g/cm^3, dC_m/dx the gradient in concentration, and D the diffusion coefficient of the diffusant in the membrane in cm^2/s.

When Henry's law applies, as with diffusing gases, the concentration of the dissolved gas is proportional to the pressure,

$$C_m = kp \qquad (8)$$

where C_m is the concentration of the gaseous diffusant, K is Henry's law solubility constant, and p is pressure.

A schematic diagram of the concentration gradient across a three-layer laminate is illustrated in Fig. 4. For purposes of this illustration, it has been assumed that the distribution coefficient is less than unity for barrier membrane I and greater than unity for barrier membrane II.

The concentration just inside the membrane surface (C_m) can be related to the concentration in the reservoir $(C_{(0)})$ by the expressions

$$C_{m(0)} = KC_{(0)} \text{ at the upstream surface } (x = 0)$$
$$C_{m(l)} = KC_{(l)} \text{ at the downstream surface } (x = l) \qquad (9)$$

where K is a distribution coefficient analogous to the familiar liquid–liquid partition coefficient.

In the following discussion, diffusion coefficients and distribution coefficients are assumed to be constant. This is a safe assumption for most polymer-diffusant

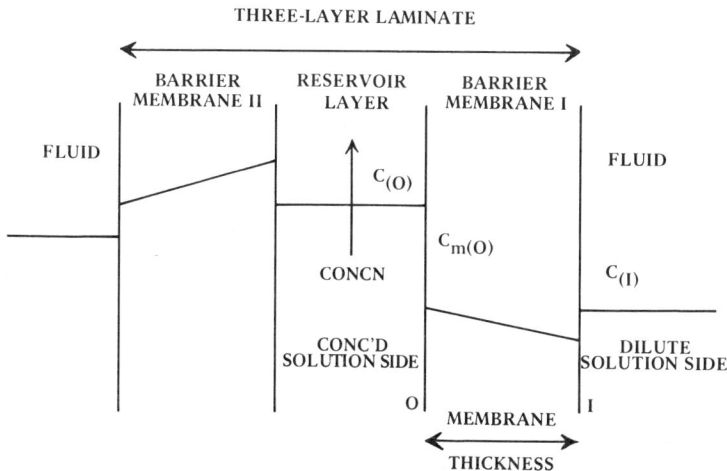

FIGURE 4. Concentration gradient across a three-layer laminated dispenser.

(permeant) systems. Thus, in the steady state or when there is a constant gradient across the membrane, Eq. 7 can be integrated to give

$$J = D\frac{C_{m(0)} - C_{m(l)}}{l} = D\frac{\Delta C_m}{l} \tag{10}$$

where l is the thickness of the membrane. Since the concentration within the membrane is usually not known, Eq. 10 is frequently written as

$$J = \frac{dM_t}{dt} = \frac{ADK \, \Delta C}{l} \tag{11}$$

where M_t is the mass of agent released, dm_t/dt is the steady-state release rate at time t, A is the surface area of the barrier membrane, DK is the membrane permeability, and ΔC is the difference in concentration adjacent to the barrier membrane.

It is important to note that the rate of release is proportional to diffusivity D (a kinetic constant) and to the distribution coefficient K (a thermodynamic constant). Equation 11 can be integrated between the limits,

$$M_t = 0 \qquad t = 0$$
$$M_t = M_t \qquad t = t$$

to give

$$M_t = \frac{ADK \, \Delta C}{l} t \tag{12}$$

2.4.1. Reservoir Layer Chemically Dissimilar to Outer Control Layers

When the distribution coefficient between the reservoir layer and the barrier membrane is much smaller than unity, as in the case of membrane I in Fig. 4, the system will have excellent release kinetics, and the release rate can be maintained constant for extended periods of time. This situation is known as pseudo-zero-order delivery. Equation 12 governs the process, and when M_t, the mass of agent released, is plotted against time t, a straight line is obtained.

2.4.2. Reservoir Layer Chemically Similar to Outer Control Layers

When the distribution coefficient between the reservoir layer and the barrier membrane is approximately unity, or larger than unity, as is the case with membrane II of Fig. 4, the Hercon system will approximate the "monolithic system," that is, the reservoir–membrane system forms a single homogeneous polymeric film. The concentration in the reservoir will not remain constant but will fall continuously with time. When the reservoir nears depletion, the system remains continuously under unsteady-state conditions and the mass of agent released varies as a function of time. This is described as "first-order delivery."

The transport equations have been described by several investigators.[148-152] Two useful equations are the *early time approximation*,

$$\frac{M_t}{M_\infty} = 4 \left[\frac{D_t}{\pi l^2}\right]^{1/2} \qquad 0 \leqslant \frac{M_t}{M_\infty} \leqslant 0.6 \qquad (13)$$

which holds over the initial portion of the release curve, and the *late time approximation*,

$$\frac{M_t}{M_\infty} = 1 - \frac{8}{\pi^2} \exp\left[\frac{-\pi^2 D t}{l^2}\right] \qquad 0.4 \leqslant \frac{M_t}{M_\infty} \leqslant 1.0 \qquad (14)$$

Equation 13 suggests that a plot of M_t vs. t will give a parabolic curve. This was found to be the case for the PVC-PVC system that dispensed the MGK R-874 fly repellent[153] (Fig. 5). When the same release rate data are plotted against $t^{1/2}$ (Fig. 6), a linear curve is obtained [153] in accordance with Eq. 13.

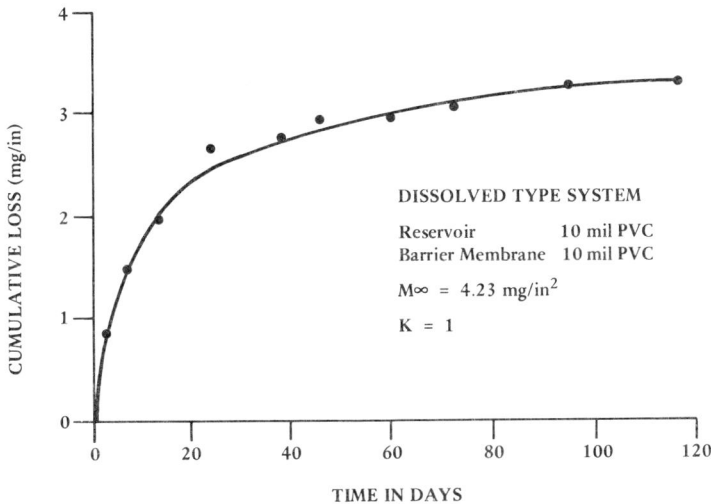

DISSOLVED TYPE SYSTEM

Reservoir   10 mil PVC
Barrier Membrane   10 mil PVC

$M\infty = 4.23$ mg/in^2

$K = 1$

TIME IN DAYS
FIGURE 5.   Parabolic release rate.

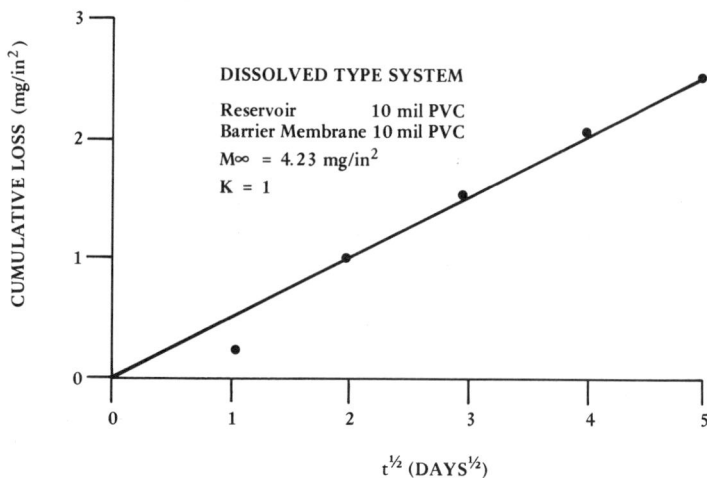

FIGURE 6. Linear release rate.

2.4.2. Factors Affecting Release Rates

Molecular and structural factors control the release of active ingredients from Hercon laminated controlled-release dispensers, as they do with other devices. For a given combination of polymer structure and active agent where energy to free rotations, free volume, and intermolecular attractions are constant, two parameters that play an important role in regulating the rate of transfer are *reservoir concentration* and *membrane thickness.* Their effect is quantified in Eqs. 12-14.

Other related factors affecting the transport of active ingredients through Hercon membranes include *polymer stiffness, codiffusants, molecular weight of diffusant,* and *chemical functionality.*

Diffusivity D and reservoir/membrane distribution coefficient K are directly proportional to the permeation rate. In polymers, diffusivity is strongly influenced by the molecular weight of the diffusant and by the stiffness of the backbone of the polymeric membrane. Simply speaking, the diffusant molecule will have to reorient several segments of polymer chain to allow its passage from site to site. The higher the molecular weight, the more the segments have to be reoriented to permit passage; and the stiffer the polymer (such as those that are glassy or highly crystalline), the more difficult it is for the segments to undergo large reorientations. Thus, variables or additives that affect polymer membrane stiffness, such as codiffusants that soften, plasticize, or partially dissolve the membrane, will also affect diffusivity and permeation rate.

The reservoir/membrane coefficient can be estimated from the solubility parameter of the diffusant, which can be calculated using Hildebrand's solubility theory. Solubility parameters and dissolution are strongly affected by molecular weight and the chemical functionality of the molecule, that is, its hydrogen bonding and polarity. When the solubility parameters for the diffusant and polymer membrane are similar, the polymer will be soluble in the diffusant. "Like dissolves like" is still a good rule of thumb.

2.4.4. Specific Applications of Laminated Controlled-Release Technology

The Hercon dispensing system is used in a number of products that are now either marketed or in advanced stages of product development:[153-155] Herculite Staph-Chek®, Insectape® or Roach-Tape® insecticidal strips, Luretape™ pheromone dispensers for use in monitoring, trapping, and mating suppression of insects, Scentstrip™ air fresheners, granules for soil insects, and fabrics for protecting stored products from insects.

Advantages and benefits derived from multilayered laminated systems are described in the following discussion of specific controlled-release commercial and developmental products.

A. Antibacterial Fabric. Herculite Products, Inc., the parent firm of the Hercon Products Group, introduced its controlled-release technology when Staph-Chek antibacterial nylon-reinforced vinyl fabrics were introduced in 1962. The products have an inner reservoir layer that contains two fungicides, which give the products their antibacterial properties. With the protected reservoir system, mattress coverings made of Staph-Check fabrics effectively control *Staphylococcus* and *Klebsiella* bacteria throughout the life of the fabric.

Thus, Staph-Check helps provide protection from cross infections. After acceptance in hospitals as mattress tickings, Staph-Check fabrics were also used for pillow covers, cubicle and shower curtains, and wall and floor coverings in burn-treatment rooms.

The addition of an antistatic agent to the Staph-Chek product offers an added dimension to its utility. Staph-Check Anstat™ fabrics reduce explosion hazards by preventing the buildup or retention of dangerous electrostatic charges.

The multilayered antibacterial fabrics were evaluated to determine their bacteriostatic activity against gram-positive *Staphylococcus* and gram-negative *Klebsiella*.[155] As shown in Table 5, bacterial counts were reduced in Staph-Chek fabrics and in a white bacteriostatic vinyl covering prepared by the Hercon process, while counts remained unchecked in the inert control. Both *Staphylococcus* and *Klebsiella* species failed to grow on the treated multilayered fabrics (over 90% reduction compared to control). The tests also showed that cultures of bacteria such as *Brevi-bacterium ammoniagenes, streptococcus pyogenes, Alcaligenes fecalis, Escherichia coli,* and *Proteus mirabilis* failed to grow on Staph-Check material.[155]

TABLE 5. Bacteriostatic Evaluation of Staph-Chek Multilayered Antibacterial Fabrics: Product Efficacy as Measured by Percent Reduction in Bacterial Count of *Staphylococcus aureus* and *Klebsiella pneumoniae* Cultures

Sample	Bacterium	Initial Inoculation	Bacterial Count	Reduction over Inert Control (%)
Inert control	*S. aureus*	212,000	6,675,000	—
Staph-Chek fabric	*S. aureus*	212,000	9,500	99.9
Inert control	*K. pneumoniae*	156,000	4,080,000	—
Antibacterial vinyl wall cover (Hercon process)	*K. pneumoniae*	156,000	1,000	99.9
Standard vinyl wall cover [a] (two samples)	*K. pneumoniae*	156,000	6,435,000 7,020,000	0 0
Inert control	*K. pneumoniae*	230,000	5,680,000	—
Staph-Chek fabric	*K. pneumoniae*	230,000	31,000	99.5

[a] Standard vinyl wall cover did not contain any antibacterial agent.

B. Insecticidal Strips for Roach Control. Hercon Insectape® insecticidal strip is a pest control product designed for use in residences and in commercial, institutional, and industrial establishments. The active agent in the reservoir of the 1-in. by 4-in. strip may be one of three insecticides long recognized as effective cockroach killers, propoxur, 2-(1-methylethoxy)phenol methylcarbamate (trade name Baygon®); Diazinon®, O,O-diethyl O-(2-isopropyl-6-methyl-4-pyrimidinyl)-phosphorothioate; and chlorpyrifos, O,O-diethyl O-(3,5,6-trichloro-2-pyridyl)-phosphorothioate (trade name Dursban®).

Advantages of the Hercon multilayered Insectape include (1) increased duration of effectiveness, which reduces the need for frequent reapplication, (2) elimination of toxic fumes through use of nonvolatile contact insecticides, (3) reduction in toxicity of active ingredient in the dispenser compared to conventionally applied insecticide, and (4) increased safety to user because toxicants are premeasured, thus eliminating spills and mixing errors.

C. Pheromone Products. The Hercon technology has also been used to dispense insect attractants or pheromones, a useful technique in pest management. Utilization of these innocuous chemical attractants signals a marked departure from sole reliance on insecticides for agricultural pest control, and interest in sex attractants as an important tool in the effort to develop alternative methods of insect pest control has been steadily mounting.[156,157] Because of the highly specific action of these compounds, damage to nontarget species can be avoided and contamination of the environment by insecticides minimized.

Initially the U.S. Department of Agriculture preferred testing a great variety of chemicals to find a synthetic insect attractant instead of trying to isolate and identify an attractive chemical from an insect. Half a dozen attractants were identified and put to good use.

Since the early 1960s, traps baited with synthetic lures for the Mediterranean fruit fly, *Ceratitis capitata* (Wiedemann); oriental fruit fly, *Dacus dorsalis* Hendel; and melon fly, *Dacus cucurbitae* Coquillett, have been deployed about ports of entry to the United States to detect any accidental importation of these serious pests. Infestations of all three species have been found, and the pests were eradicated quickly with the aid of traps which showed where and when to apply the appropriate controls. The USDA stated that although the traps saved them millions of dollars in potential eradication costs, they saved the agricultural community and the public even more in terms of the increased prices of food that would have resulted if these pests had remained unchecked.

With the great value of insect attractants well established, even greater efforts were put into the isolation, identification, and synthesis of the compounds that the insects themselves manufactured (the pheromones). At first, work in this area was hindered by the minute amounts of these potent substances present in insects. In the late 1960s, however, remarkable improvements in spectrophotometric, chromatographic, and other instrumental and specialized techniques facilitated the isolation and identification of minute amounts of pheromones. Progress has been so rapid in the past five years that pheromones or chemicals believed to be pheromones are now known for hundreds of insect species,[158] and many of these are of great economic importance.

The Luretape® pheromone dispenser is usually manufactured as a three-layer plastic strip with the insect attractant contained in the inner layer. As in other Hercon dispensers, the two outer plastic layers act to regulate the release of the pheromone. The pheromone gradually diffuses from the reservoir through the outer layers to the exposed surface, from which it evaporates into the atmosphere. At the same time, the outer layers protect the bulk of the pheromone from the degradative action of the environment. Sensitive compounds, such as the aldehydes or other unsaturated chemicals that are the pheromones or pheromonal components of some of the world's most important insect pests, are particularly vulnerable to degradation and deactivation. The Hercon multilayered laminate is the only commercially available effective dispenser of aldehydes that has been used in large-scale programs. The Hercon dispenser not only regulates release of the pheromone, but also protects the compound from degradation and oxidation by sunlight and weather.

In addition to protecting unstable and volatile chemicals and controlling their release rate, the Hercon pheromone dispenser offers economic benefits in terms of long-term effectiveness, elimination of reapplication costs, and reduction in the amount of active ingredient needed.

In an integrated pest management program, pheromones may be used in lure-baited traps to signal the presence of a specific insect pest. Their high degree of specificity makes them particularly valuable in monitoring pest populations. After the level of infestation is detected and estimated, control measures may be applied, either to the surrounding area to prevent the spread of the insect or only to those areas where the pest is found. The traps can also be used to time the application of control measures and thereby cut down on the number of treatments needed.

In addition to survey and detection, pheromones may also be used as direct control measures for (1) mating suppression through insect disorientation and (2) mass trapping. Pheromones may also be used with "trap crops." The bait (or attractant dispenser) is applied to plants (the trap crop) in a specific area, and the area is then sprayed with an insecticide after the target insects have been lured to the trap-crop area.

The trap-crop technique was evaluated by the USDA as part of the trial program in 1972 and 1973 to eradicate the cotton boll weevil, *Anthonomus grandis* Boheman.[159] A four-row strip of trap-crop cotton was planted in every cotton field under test. The trap crop was planted two to three weeks prior to planting of producer's cotton in order for the trap crop to be larger, fruit earlier, and be generally more attractive to the insect. To further enhance the attractiveness of the trap crop, dispensers of grandlure, the cotton boll weevil pheromone, were applied, either to prevent the spread of the insect to the surrounding area or weevils that aggregated on the trap crop.

Results of the pilot trap-crop experiment indicated that a combination of the bait and the systemic insecticide treatment effectively controlled overwintering boll weevils. The researchers observed that almost all the weevils that entered the field before the appearance of flower buds on the producer's cotton moved onto the trap crop and were killed after feeding on the systemic insecticide-treated plants.[160]

Table 6 lists the insects whose pheromones have been incorporated into the Hercon system.

The USDA gypsy moth survey program was the first to use the Hercon controlled-release pheromone dispensers.[204] The active ingredient for such dispensers is the sex pheromone emitted by the virgin female to attract the male, a single compound identified as *cis*-7,8-epoxy-2-methyloctadecane and called disparlure.[205] Throughout the 1970s, several hundred thousand traps were strategically placed in the eastern half of the United States to monitor the presence of the gypsy moth, the most important defoliating insect of hardwoods, especially the oak, in the northeastern states. The rate of emission of disparlure from the dispensers was found to be ample and fairly constant, and the dispensers were effective for an entire season.[205]

Mass trapping with a new, inexpensive, and highly efficient trap baited with a Luretape showed potential for direct control of the gypsy moth in light infestations.[206] Because of the success of this approach, pheromone-baited gypsy moth traps, sold under the tradename Lure N Kill[TM], are now available to the general public.

Permeating the atmosphere with disparlure released from Hercon dispensers to disorient males trying to find the nonflying female gypsy moth has resulted in as much as 98% suppression in mating.[206-208,210-212] Success in mating disruption hinges on the dispenser's ability to provide an adequate pheromone concentration in the atmosphere throughout the insect's flight period.[209] By trapping and measuring released pheromone, the emission rate of disparlure from a Hercon dispenser under controlled conditions has been shown to be relatively constant.[204,209,210] Hercon laminated $\frac{1}{8}$ in. by $\frac{1}{8}$ in. "flakes" known as Disrupt have also been applied aerially and have successfully suppressed mating. The long-lasting quality of the laminates makes possible their application well in advance of the gypsy moth flight.

The Hercon pheromone dispenser has also been used successfully by the USDA in an integrated control program against the cotton boll weevil, considered the worst pest of cotton. Luretape used in traps contained the boll weevil sex and aggregating pheromone,[214-216] a four-component mixture of chemicals,[213] two of which are aldehydes: *cis*-3,3-diemthyl-Δ^1,α-cyclohexaneacetaldehyde and *trans*-3,3-dimethyl-Δ^1,α-cyclohexaneacetaldehyde; and two are alcohols: (+)-*cis*-2-isopropenyl-1-methylcyclobutaneethanol and *cis*-3,3-dimethyl-Δ^1,β-cyclohexaneethanol.

A summary of the comparison between Luretape and two other formulations is illustrated in Fig. 7. Other formulations, such as impregnated cigarette filter in plastic tubes or gel inside a plastic straw, hollow fiber dispensers, and grandlure-containing vials with a flexible foam barrier, were not as effective as the Luretape.

The longer-lasting effectiveness of Luretape-baited traps is assured by the linear release rate of grandlure (Fig. 8). This was the result of the outer polymeric layers of the dispenser controlling the rate of release and preventing the fast degradation of the aldehyde components of the pheromone. Hercon's Environmental Protection Agency registration allows marketing of the traps, known as the Boll Weevil Scout[TM], in the United States. Marketing also continues in Central and South America.

Although grandlure has been used primarily in mass trapping, successful mating-disruption work using Luretape as the source of the pheromone has also been conducted.[217]

TABLE 6. List of Insects whose Behavior-Modifying Chemicals Have Been Incorporated into Hercon® Multilayered Controlled-Release Dispensers (as of June 1981)

Order and Family	Common Name	Scientific Name[a]	Application[b]	Ref.
Coleoptera				
Curculionidae	Boll weevil	*Anthonomus grandis* Boheman	M, MT	161–163
Dermestidae	Khapra beetle	*Trogoderma granarium* Everts		
Scarabaeidae	Japanese beetle	*Popilla japonica* Newman	M, MT	164
Scolytidae	Ambrosia beetles	*Gnathotrichus sulcatus* (Le Conte) and *G. retusus*		
Scolytidae	Smaller European elm bark beetle	*Scolytus multistriatus* Marsham	M, MT, TT	165–167
Scolytidae	Southern pine beetle	*Dendroctonus frontalis* Zimmermann		
Scolytidae	Spruce bark beetle	*Ips typographus* (L.)	MT	168, 169
Diptera				
Muscidae	House fly	*Musca domestica* L.		170, 171
Tephritidae	Mediterranean fruit fly	*Ceratitis capitata* (Wiedemann)	M	172
Homoptera				
Pseudococcidae	Comstock mealybug	*Pseudococcus comstocki* (Kuwana)		
Hymenoptera				
Vespidae	Western yellowjacket	*Vespula pensylvanica* (Saussure)		
Lepidoptera				
Gelechiidae	Angoumois grain moth	*Sitotraga cerealella* (Walker)	DM	173
Gelechiidae	Peach twig borer	*Anarsia lineatella* Zeller		
Gelechiidae	Pink bollworm	*Pectinophora gossypiella* (Saunders)	M, DM	174–176
Lymantriidae	Douglas-fir tussock moth	*Orgyia pseudotsugata* (McDunnough)	M	177
Lymantriidae	Gypsy moth	*Lymantria dispar* (L.)	M, MT, DM	178–185
Noctuidae	Black cutworm	*Agrotis ipsilon* (Hufnagel)	M	186
Noctuidae	Fall armyworm	*Spodoptera frugiperda* (J. E. Smith)	DM	187
Noctuidae	Egyptian cotton leafworm	*Spodoptera litteralis* (Boisduval)	DM	

Family	Common name	Scientific name[a]	Legend	Reference
Noctuidae	Cabbage looper	Trichoplusia ni (Huber)	M	188
Noctuidae	Soybean looper	Pseudoplusia includens (Walker)	M	
Noctuidae	Bollworm/corn earworm/ tomato fruitworm	Heliothis zea (Boddie)	M	189, 190
Noctuidae	Tobacco budworm	Heliothis virescens (Fabricius)	M	172, 186, 190–194
Noctuidae	American (= old world) bollworm	Heliothis armigera (Hubner)	M	
Olethreutidae	Codling moth	Laspereysia pomonella (L.)	M, DM	
Olethreutidae	Oriental fruit moth	Grapholitha molesta (Busck)	DM	195, 196
Olethreutidae	Western pineshoot borer	Eucosma sonomana Kearfott	DM	197
Pterophoridae	Artichoke plume moth	Platyptilia carduidactyla (Riley)	M	
Pyralidae	Asiatic rice borer	Chilo suppressalis (Walker)	DM	
Pyralidae	Indian meal moth	Plodia interpunctella (Hubner)	DM	
Pyralidae	Naval orangeworm	Amyelois transitella (Walker)	DM	198
Sesiidae	Grape root borer	Vitacea polistiormis (Harris)	DM	199
Sesiidae	Lesser peachtree borer	Synanthedon pictipes (Grote and Robinson)	DM	200–202
Sesiidae	Peachtree borer	Synanthedon exitiosa (Bay)	DM	200–202
Tortricidae	European grape(vine) moth	Eupoecilia ambiguella Hubner	DM	203
Tortricidae	Spruce budworm	Choristoneura fumiferana (Clemens)	DM	
Tortricidae	Western spruce budworm	Choristoneura occidentalis Freeman	DM	

[a] Scientific names used are from Common Names of Insects and Related Organisms[248].

[b] Unless specified, Hercon dispensers of specific pheromones or attractants were formulated for preliminary evaluations. Legend: M represents monitoring or survey; MT, mass trapping; DM, disruption of mating or confusion technique; and TT, trap tree system.

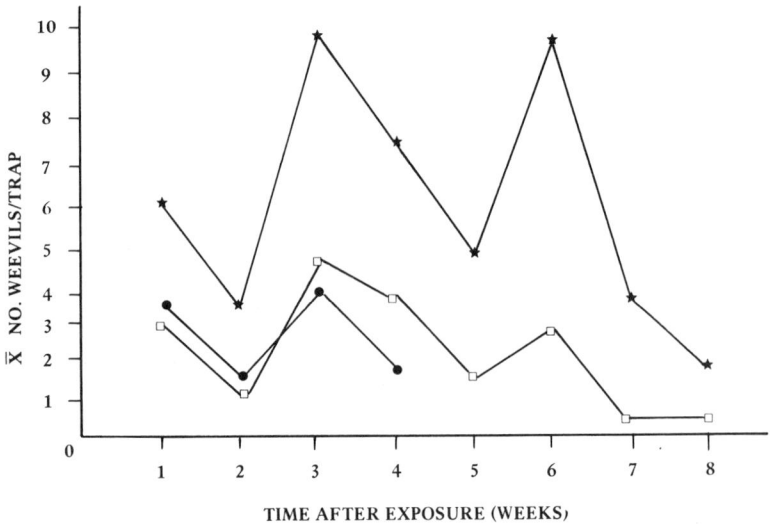

FIGURE 7. Comparison of boll weevil trap catches using various controlled-release devices.[163] ★, 3-layer polymeric laminate (Hercon® dispenser); ●, hollow fiber dispenser (Conrel®); □, polyethylene glycol mixture impregnated in a cigarette filter enclosed in a glass vial.

Hercon dispensers of the pink bollworm gossyplure pheromone have been used in the USDA survey programs. The baits, 1 cm² size and containing about 1.5 mg of the lure a 1:1 mixture of *cis,cis* and *cis,transisomers* of 7,11-hexadecadien-1-ol acetates, have been highly effective in capturing male pink bollworm moths during field trials in different parts of the country.

Table 7 compares the number of moths caught in traps baited with Luretape and with cotton dental roll wick containing the same lure.[218] The data show that the Hercon formulations were more effective than the cotton wicks in luring the moths to the traps and had a longer duration of effectiveness.

In 1979, the Western Cotton Research Laboratory at Phoenix, Arizona, undertook an extensive experiment in which cotton was aerially treated with Hercon flakes containing gossyplure to determine the effectiveness of this treatment in disrupting the mating of the pink bollworm.[175] The results were excellent and provided the first scientific evidence of the success of this technique for the pink bollworm.

The smaller European elm bark beetle has been responsible for the decline of the American elm tree, *Ulmus americana* L. It is the major vector of the virulent and destructive parasitic fungus, *Ceratostomella ulmi* Schwarz, of the Dutch elm disease. After the ingredients of the beetle's multicomponent pheromone were identified, these chemicals were formulated into Hercon dispensers, which

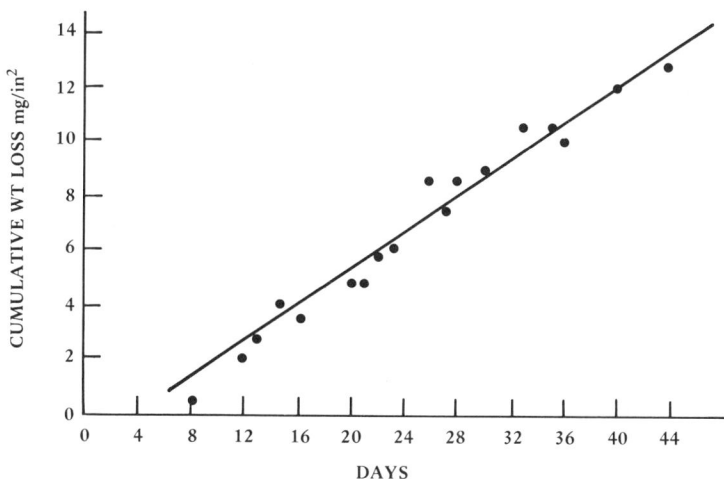

FIGURE 8.   Release rate of grandlure from a Hercon® laminated dispenser.[161]

proved to be highly attractive to the insects.[219,220] These results indicate that only one rebaiting would be needed to retain trap effectiveness for the entire 150—180 day-season of the beetle.

Hercon dispensers containing *cis*-8-dodecenyl acetate, the synthetic sec pheromone of *Grapholita molesta* (the Oriental fruit moth), were compared with NCR plastic-coated gelatin-based microcapsules to evaluate their efficiency in disorienting the male moths seeking females for mating.[195] Tests were conducted from 1973 to 1975 at the Southeastern Fruit and Tree Nut Research Station in Byron, Georgia. The dispensers, one on every other tree of a 2-acre peach orchard, at a height of 1.2 m, suppressed response of the male moths 100% to traps baited with virgin females for 12 weeks. Twig damage was reduced from 78 to 100% (compared to an untreated orchard) during the same 12-week period, and the lure in the strips decreased from 2.26 to 1.17%. The microencapsulated formulations were effective for only 5 weeks.

The peachtree borer and the lesser peachtree borer have been successfully disoriented in peach orchards.[222,222] In an experiment conducted by C. R. Gentry and associates, 1 in. by 1 in. Luretape dispensers of *cis,cis*-3,13-octadecadien-1-ol acetate, the peachtree borer pheromone, were hung on trees in two peach orchards; another orchard was left untreated as a check. In comparison with the check, captures of the lesser peachtree borer in the pheromone-baited traps in orchards with the dispensers were suppressed 100% for 9 weeks; then the results became somewhat erratic. Fresh dispensers restored almost complete suppression of captures for another 10 weeks (Table 8). Similar results were recorded for the peachtree borer, but significance of the results is questionable because the insect population was extremely low.

TABLE 7. Comparison of Male Pink Bollworm Moths Caught in Five Traps Baited with Standard Cotton Wick and Hercon Luretape Dispensers of Gossyplure, the Pink Bollworm Pheromone [218]

| | Mean Number of Moths Caught at Specified Days | | | | | | |
Dispenser	3	7	13	22	27	34	Cumulative Number Caught
Standard cotton wick	3.0	11.4	11.6	2.8	2.0	7.0	189
Luretape 1	8.0	70.0	35.8	26.4	4.8	19.6	823
Luretape 2	8.6	90.0	54.4	36.2	7.8	29.0	1130
Luretape 3	10.2	86.4	48.2	36.6	7.2	28.8	1080
Luretape 4	20.6	36.4	54.2	25.6	1.2	32.2	911

The superiority of the Hercon dispenser in mass trapping Japanese beetles with use of a minimum amount of pheromone has been demonstrated.[164,223,224] As a result, Hercon is planning to market traps for this insect as well.

The governments of Norway and Sweden, working in cooperation, and the Borregaard AS company, have initiated what is probably the largest scale pheromone experiment in history for controlling an insect pest. Following extensive tests with several different types of dispensers containing the multicomponent pheromone of the spruce bark beetle, a killer of the valuable Norwegian spruce tree, several hundred thousand Hercon dispensers have been selected for continued use. The project was adjudged highly successful, and it is believed that several million acres of forest have been saved to date. This work is continuing.

The Hercon dispenser has also been successfully evaluated against the tobacco budworm, *Heliothis virescens* (Fabricius), the fall army worm *Spodoptera frugiperda* (J. E. Smith), the eastern spruce budworm *Choristoneura fumiferana* (Clemans). The performance of the dispensers against these insects is cited in the literature.[225,226]

With pheromones of economically important insect species rapidly becoming available, the Hercon dispensing system appears to have excellent potential for use in trapping, in air permeation, and in lure–insecticide combinations. The flexibility of the system makes it easy to adjust the rate of emission of the behavior chemical to values approaching optimum, and greatly facilitates use as baits in traps or as evaporators of the sex attractant in the air-permeation mating-suppression technique. Ease of handling and storage, protection of a broad variety of chemicals from degradation by light and air, and excellent weathering qualities are other unique features of the Hercon multilayered polymeric dispensers.

TABLE 8. Effect of Air-Permeation Trails on Captures of Lesser Peachtree Borer Males in Pheromone Traps[222]

| Week | Number of Lesser Peachtree Borers Captured[36] | | |
	Orchard 1 (Treated)[a,b]	Orchard 2 (Treated)	Orchard 3 (Untreated)
1	1 (86)	0 (100)	7
2	0 (100)	2 (93)	27
3	2 (92)	0 (100)	24
4	1 (93)	0 (100)	15
5	0 (100)	0 (100)	2
6	1 (90)	9 (100)	10
7	9 (100)	1 (92)	12
8	0 (100)	0 (100)	4
9	0 —	0 —	0
10	0 (100)	2 (83)	12

[a]Each orchard had 5 traps baited with 100 mg of pure E,Z.

[b]Value in parentheses is percent reduction compared to the untreated orchard.

D. Fragrance Dispensers. In addition to the multilayered products for pest control, Hercon has employed its patented controlled-release system to dispense fragrances, deodorizers, and malodor counteractants. Called the Scentstrip, this basic Hercon dispenser contains the scent as the active ingredient in the inner reservoir layer. The product is now commercially available and competes in the varied deodorant–air freshener market against solid gels, wicks, aerosol sprays, saturated products, and deodorizing disinfectants. The Hercon fragrance dispenser is produced in flat continuous sheets up to 60 in. wide, making it adaptable to many die-cut forms and inexpensive packaging. The dispensers may be produced in strips, squares, ovals, stars, or other shapes such as pine trees, flowers, fruits, animals, or cartoons. The strips may also be packaged singly, or they can be joined on a common backing sheet.

Scentstrip can be manufactured so that one or both sides of the strip are active in dispensing fragrance molecules. Typically, such units are designed for hanging, insertion into air-conditioning units, or inclusion in a perforated package. Dispensers with one side impervious to chemical passage can be given a coating of pressure-sensitive adhesive for easy stick-on use. In this form, the dispenser is affixed to unobtrusive surfaces such as the underside of tables, the back of picture frames, the inside surface of drawers or clothes hampers, garbage can lids, or near any other source of offensive odors.

E. Stored-Food Insect Control. Multilayered Hercon dispensers have been evaluated for the protection of stored food products. Such protection will help increase the food supply, which is dwindling as the world's population is rapidly out-pacing agricultural production.

Protection of stored food products from deterioration is made difficult by many interacting physical, chemical, and biological variables. With stored grain, the quality of the product is affected by temperature; moisture; oxygen; local climate; granary structure; and physical, chemical, and biological properties of grain bulks; as well as attack by microorganisms, insects, mites, rodents, and birds.[227] These variables seldom act alone or all at once, but their presence can significantly diminish the food supply. Controlling insect pests will not necessarily win the war against the deterioration of stored products. However, it will be a significant victory, because grain injury and organic litter due to insect feeding will be minimal, and much of the growth of fungi and bacteria can then be checked.

Fumigants and a few residual insecticides have been used widely to control general insect infestations.[228] However, several of these chemicals are encountering regulatory difficulties and may be withdrawn from use because they are potentially hazardous. Other problems, such as off-flavors, pesticide residues, and development of insecticide resistance among stored-product insects, also exist, and they have spurred the search for improved or new pest control and methods.

Fabrics with insecticides incorporated into and protected by the Hercon multilayered structure have been evaluated to determine their potential use as toxicants, repellents, or attractants against stored-product insects.[229] Such fabrics are potentially useful as packaging containers or insect-resistant barriers for stored food products. A film containing pyrethrins and piperonyl butoxide is now used successfully to protect packaged dried fruits.[230]

The USDA Stored-Product Insects Research and Development Laboratory at Savannah, Georgia, has tested various Hercon multilayered fabric formulations containing the insecticide in the inner reservoir layer. The results are presented in Tables 9 and 10. Residual toxicity and repellency–attractancy tests were also conducted.[231,232]

F. Polymeric Granules containing Conventional Agrichemicals. Hercon granular formulations, which are ground-up laminated material, have been evaluated for their effectiveness against soil insects and for other agricultural and turf applications. These experimental products have been tested against the northern corn rootworm, *Diabrotica longicornis* (Say); banded cucumber beetle larvae,

TABLE 9. Percent Mortality of Adult Confused Flour Beetles, *Tribolium confusum*, Five Days After 24 hr Exposure to the Surfaces of Hercon Insecticidal Dispensers[231]

Insecticide	Content (Wt%)	Date Fabricated (1972)	Mortality (%) at Indicated Test Period (Date)					
			0 (1/2/73)	6 mo (7/10/73)	1 yr (1/12/74)	2 yr (1/7/75)	3 yr (1/6/76)	4 yr (1/5/77)
Carbaryl	15	11-28	0	0	10	0	3	0
Chlordane	5	8-19	83	80	67	37	0	3
	19	11-28	97	100	100	97	23	24
Chlorpyrifos	1	8-19	50	40	63	37	53	38
	15	11-28	100	100	100	100	100	100
Stirofos	8.5	11-28	100	100	100	100	100	100
Malathion	5	4-15	97	100	100	100	93	100
	16	11-28	100	100	100	100	100	100
Methoxychlor	5	4-15	7	3	0	0	0	0
Pyrenone	5	4-15	3	0	10	0	0	0
Control	0		0	0	0	0	0	0

TABLE 10. Percent Mortality of Black Carpet Beetle Larvae, *Attagenus megatoma*, 14 days After a 24-h Exposure to the Surfaces of Hercon Insecticidal Dispensers[231]

Insecticide	Content (Wt%)	Date Fabricated (1972)	Mortality (%) at Indicated Test Period (Date)							
			0 (1/2/73)	6 mo (7/10/73)	1 yr (1/12/74)	2 yr (1/7/75)	3 yr (1/6/76)	4 yr (1/5/77)
Carbaryl	15	11-28	0	3	20	0	0	2
Chlordane	5	8-19	0	0	0	0	0	0
	19	11-28	17	17	87	0	3	3
Chlorpyrifos	1	8-19	93	90	100	93	63	92
	15	11-28	100	97	100	97	90	100
Stirofos	8.5	11.28	27	97	93	100	77	96
Malathion	5	4-15	100	100	97	97	73	99
	16	11-28	100	100	97	97	97	99
Methoxychlor	5	4-15	0	3	0	0	0	0
Pyrenone	5	4-15	0	0	0	0	0	0
Control	0		0	0	0	0	0	0

D balteata LeConte; southern corn rootworm, *D. undecimpunctata howardi* Barber; western potato leafhopper, *Empoasca abrupta* DeLong; cabbage maggot, *Hylemya brassicae* (Bouche); sweet corn borer, *Sesamia nonagrioides;* and white grub larvae of European chafer, *Amphimallon majalis* (Razoumowski) and Japanese beetle, *Popillia japonica* Newman, among other common insect pests.

Protecting the active ingredient under field conditions is necessary when the local environment adversely affects the toxicant's stability. Thus, some insecticides are superior as foliar sprays but are ineffective below ground against soil insects. Conventional granular formulations are also liable to be highly susceptible to hydrolysis, photodecomposition, or volatilization and therefore ineffective in protecting the active ingredient from rapid loss of effectiveness.

Controlled-release products generally offer definite economic advantages by prolonging the effectiveness of insecticides, eliminating costly overspraying, and reducing the toxic hazard of the compounds to the applicator by sealing them within a protective structure. These long-lasting products are also beneficial because they reduce or eliminate reapplications.

Hercon controlled-release granules containing 8.3 and 11.3% diazinon were used to protect the roots of field corn from rootworm feeding.[233] Tests showed that the plants treated with the Hercon 8.3% formulation had the least root damage and highest yields per acre. Plants receiving the commercial Diazinon® 14G formulation had roots damaged by larval feeding.

Against *D. balteata* larvae, Hercon 6% diazinon granules outperformed the standard Diazinon 14G product in terms of percent larval mortality and duration of effectiveness[234] (Table 11). In laboratory tests, soil treated with Hercon granules (2 ppm) gave 80% larval mortality 13 weeks after treatment, while Diazinon 14G (2 ppm) gave 50% mortality or less after only 9 weeks. The 1-ppm rates of the Hercon and standard products showed 60% mortality after 9 and 7 weeks, respectively.

The results demonstrated that the Hercon granules protected young corn plants from injury by soil insects for at least 13 weeks, yet the insecticide concentration was substantially lower than that present in commercially available products providing the same duration of effectiveness.

In a greenhouse study, Hercon granular formulations of Thimet® systemic insecticide were compared with the standard Thimet clay granules for effectiveness against the southern corn rootworm.[234] As shown in Table 12, the standard attapulgite clay granules reduced or eliminated feeding damage only after applying at least 30 mg of the toxicant. At lower rates (7.5 and 15 mg a.i.), effectiveness lasted only 73 days, and severe feeding damage was observed 94 days after treatment. The Hercon formulation protected the plants throughout the 164-day study at all four rates.

Both the southern corn rootworm and leafhopper studies showed that the Hercon formulation can provide the same degree of effectiveness as the standard

TABLE 11. Percent Mortality of Banded Cucumber Beetle Larvae, *Diabrotica balteata*, Held in Field Soil Treated with Diazinon in Hercon and Commercial Granular Formulations[234]

		Percent Mortality at Indicated Week							
Treatment	Rate (ppm)	1 Week	3 Week	5 Week	7 Week	9 Week	11 Week	13 Week	15 Week
Diazinon-Hercon	1	90	100	90	70	75	60	45	0
6.0%	2	75	100	90	100	80	95	80	35
	4	85	95	100	100	100	95	100	95
Diazinon-14G	1	75	90	80	60	30	25	25	0
	2	90	100	100	95	50	30	20	10
	4	100	85	70	95	100	85	90	30
Control 1	—	15	35	0	20	25	25	30	0
Control 2	—	10	25	15	15	5	25	20	5

TABLE 12. Effectiveness of Thimet® Insecticide in Hercon and Attapulgite Clay Granular Formations Against the Southern Corn Rootworm, *Diabrotica undecimpunctata howardi*[234]

Treatment System	Rate AI Thimet (mg)	Average Corn Damage Rating[a] at Specified Days After Treatment					
		17	42	73	94	122	164
Hercon	—	3.0	3.0	3.0	3.0	3.0	3.0
Attafulgite	7.5	1.0	0.7	0.3	3.0	3.0	3.0
clay	15.0	1.0	0.0	0.0	2.3	3.0	3.0
	30.0	0.0	0.0	0.0	0.0	0.7	1.7
	60.0	0.0	0.0	0.0	0.7	0.0	0.0
Hercon	7.5	0.7	0.3	0.0	1.0	0.3	0.0
	15.0	0.7	0.0	0.0	0.0	0.0	0.0
	30.0	1.0	0.0	0.0	1.3	0.7	0.0
	60.0	0.0	0.0	0.0	0.0	0.0	0.0
Control	—	3.0	3.0	3.0	3.0	3.0	3.0

[a] Feeding damage rating: 0, no feeding; 1, slight feeding; 2, moderate feeding; 3, severe feeding.

while using only one-fourth to one-eighth the amount of Thimet. This is added evidence that effectively protecting the active ingredient from degradation can reduce the quantity of toxicant used without sacrificing plant protection and long-lasting performance.

Hercon granular formulations of diazinon and chlorpyrifos were tested for cabbage root maggot control following in-furrow application.[233] Roots of plants treated with Hercon 8.3 and 11.3% diazinon granules showed only 16.7 and 20% damage, respectively, while those treated with the commercial Diazinon 14G product had 53.3% damage. Roots of the untreated check had 60% damage. The same study also evaluated spray and granular insecticidal materials for control of European chafer and Japanese beetle larvae.

Under a recent grant from USDA, Hercon has also been developing a slow-release insecticide product for use against the imported fire ant (*Solenopsis invicta* Buren) in the nursery and sod industry. Amaze®, chlorpyrifos, and diazinon were chosen as the insecticides to be tested.

Twenty-four Hercon formulations were evaluated by bioassays of the treated soil. Insecticide concentrations were also determined. At 11 months posttreatment, the chlorpyrifos formulations were found to be by far the most effective, based on chemical and biological data. Release rates from the Hercon granules in potting soil indicate that the 20% chlorpyrifos (initial concentration) granules are likely to remain effective for the desired 2-year period. The study will be continued for another year to determine which specific formulations are best and their duration of effectiveness.[235]

Effectiveness of insecticide granules, whether conventional or encapsulated, depends on a number of factors including insect species, host crop, and release rates. A "quick-release" granular formulation may be satisfactory against some insects, but the sustained "slow-release" types may be better if a longer duration of effectiveness is desired or if an easily degraded toxicant is used.

While controlled-release formulations will definitely play an important role in agricultural pest management, in some situations this technology may be impractical and expensive. Each case must be considered individually, and for those controlled-release products that turn out to be higher in cost, this cost must be weighed against desired benefits.

2.5. Other Physical Methods

2.5.2. Osmotic Pumps

Osmotic pumps comprise another system that is both novel and useful. In its simplest and most elegant design, the pump consists of a tablet containing the active agent and an osmotic "attractant," such as NaCl, surrounded by a semipermeable membrane with a small orifice. When the pump is placed in an aqueous environment, the osmotic pressure of the NaCl draws water to the device through the semipermeable membrane. Because the membrane coating is nonextensible, the NaCl-saturated solution (and active agent) inside the device is pumped out through the orifice as water is imbibed osmotically.

The hydrostatic pressure difference $p_2^H - p_1^H$, which is the driving force for the flow of water across the semipermeable membrane, is given by the equation

$$p_2^H - p_1^H = \pi_2 - \pi_1 = \frac{RT}{V} \ln \frac{P_2^v}{P_1^v} \tag{15}$$

where R is the gas constant, T the absolute temperature, V the molar volume of the solvent (water), P_1^v and P_2^v the vapor pressure of solvent above solutions 1 and 2, and $\pi_2 - \pi_1$ the osmotic pressure at equilibrium.

The volume flow dV/dt of water across the semipermeable membrane when going toward equilibrium is given by the equation[236]

$$\frac{dV}{dt} = \frac{A}{h} L_p [\sigma(\pi_2 - \pi_1) - (p_2^H - p_1^H)] \tag{16}$$

where A is the membrane area, h the membrane thickness, L_p the mechanical permeability coefficient, and σ the reflection coefficient characterizing the leakiness of the membrane.

As can be seen from Eq. 16, the amount of water introduced into the osmotic pump remains constant with time. Thus the delivery rate of the solute by the pump is constant (zero-order release) as long as excess solid is present inside the pump to form a saturated solution. The rate declines exponentially (first-order) as soon as the solution inside the device drops below saturation.

Early development systems[237-239] used Congo red as the osmotic driving agent and cellophane as the semipermeable membrane. These devices delivered material for as long as 100 days at low rates and for a few days at as much as 0.5 ml/day.

During the last 10 years, the Alza Corporation has developed and patented a series of sophisticated osmotic controlled-release systems under U.S. Patents 4034758, 4036227, 4036228, 4058122, and 4077407. The first two patents pertain to delivery of drugs to uterus, eye, and stomach, and the last three to delivery of orally administered medicaments such as antacids, analgesics, and vitamins.

The basic osmotic unit for oral delivery is the elementary osmotic pump[240,241] shown in Fig. 9. It consists of an active agent admixed with an osmotic driving agent as the solid core, surrounded by a semipermeable membrane containing a delivery orifice. The system can be made of any specified size or shape. The delivery rate is programmed by the osmotic pressure of the core formulation and by the permeability of the semipermeable membrane to water.

Another osmotic unit is the miniosmotic pump.[242] The minipump consists of an inner collapsible reservoir surrounded by osmotic driving agent, potassium sulfate. The driving agent, in turn, is surrounded by a cellulosic semipermeable membrane. The system is fabricated and sold unfilled and programmed at a volume delivery rate fixed by selection of the osmotic driving agent and the membrane.

Osmotic delivery systems are well suited for applications where the system can be preprogrammed to deliver the agent at a specified rate and water is available at sufficiently high activity. These include pharmaceutical applications for delivery of herbicides, fungicides, and insecticides. The cost of preparing the osmotic pressure devices is the biggest shortcoming, and usually less sophisticated controlled-release systems are found to be adequate for most applications.

2.5.2. Gels

Gels are jellylike materials formed from the precipitation of sols, which are fluid colloidal solutions. Colloidal solutions are systems that consist of media with dissolved or dispersed particles ranging in diameter from approximately

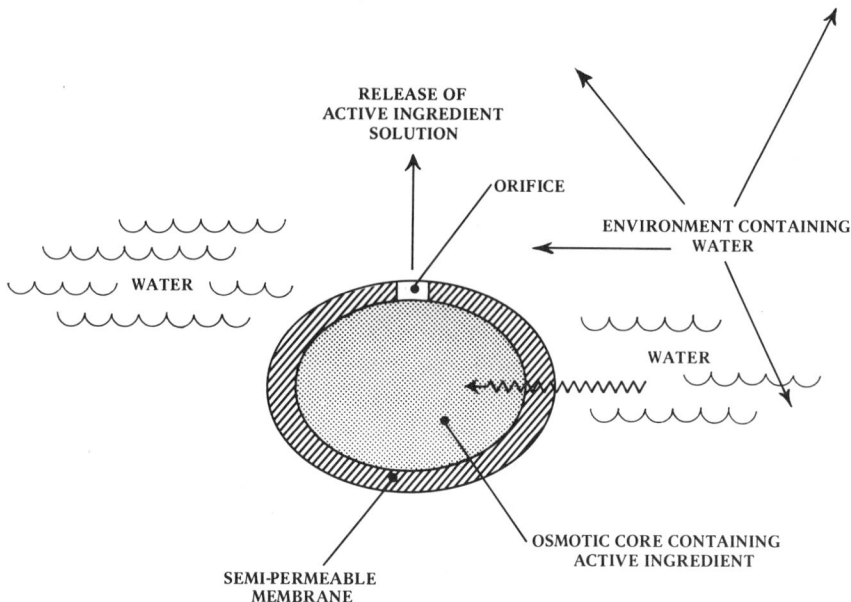

FIGURE 9.   Cross section of an osmotic pump.

1 mm to 1 μm. The coagulation and precipitation of sols can be affected by the addition of certain ions to neutralize the electric charges, boiling, freezing, and the addition of solvents with small amounts of an electrolyte. The precipitate formed may or may not be gelatinous. However, if conditions are right, it is possible to obtain the dispersed phase as a more or less rigid mass enclosing all the liquid within it.

In controlled-release technology, gels are used almost exclusively for air fresheners for the consumer market. Carrageenan is used as the gelling agent, with smaller amounts of locust bean gum, agar-agar, and other water-soluble polymers also being used. The total amount of the gelling agent is 1.5-2.5% of the total weight of the gel. The dispersion medium is between 85 and 95% water, with the remainder being ethyl alcohol, ethylene glycol, a surfactant, a dye, and 2-4% perfume.[243] The gels are prepared by dispersing the carrageenan and other ingredients in the medium, heating the solution to 180°F to effect dissolution, and then cooling to room temperature to provide the air freshener gel.

Recently, gels have been prepared using the cross-linking of macromolecules in solution.[244,245] The formulation is similar to the one mentioned above except that the gelling agent is 3% sodium cellulose cross-linked with basic chromic sulfate. The major advantages of these gels are that gelling is instantaneous, and no heating and cooling are required.

One may consider gels as a special case of monolithic devices. However, gels differ from other devices of this type by the substantial change in the size of the gel itself as its contents are released. In addition to the complexity mentioned above, the air freshener gels release their contents in substantially linear fashion.[246]

2.5.3. Adsorption onto Ion-Exchange Resins

Adsorption of active agents onto ion-exchange resins has been tried as a controlled release mechanism. [247] Thus the adsorptive forces of these resins can decrease the release of an ionic species through an equilibrium favoring the resin's adsorption sites, but renewal of the medium can result in very fast release. In pharmaceutical applications, the release rate can depend upon pH and electrolyte concentration. For example, in the gastrointestinal tract, release can be higher in the stomach and lower during transit through the small intestine, owing to pH differences.[248]

REFERENCES

1. S. K. Chandrasekaran, "Theory of Controlled Delivery Systems," in R. L. Goulding, Ed., *Proc. Int. Controlled Release Pesticide Symp. Oregon State University, Corvallis,* 1977, p. 382.

2. J. A. Bakan, "Microencapsulation of Pesticides and Other Agricultural Materials," in F. W. Harris, Ed., *Proc. Int. Controlled Release Pesticide Symp., Wright State University, Dayton, OH,* 1975, p. 76.

3. G. O. Fanger, "General Background and History of Controlled Release," in N. F. Cardarelli, Ed., *Proc. Int. Controlled Release Pesticide Symp., University of Akron, Akron, OH,* 1974, p. 1.18.

4. H. B. Scher, "Microencapsulated Pesticides," in H. B. Scher, Ed., *Controlled Release Pesticides (ACS Symp. Ser. 53),* Am. Chem. Soc., Washington, DC, 1977, p. 127.

5. J. R. Robinson, "Controlled Release Pharmaceutical Systems," in F. W. Long, W. P. O'Neill, and R. D. Stewart, Eds., *Chemical Marketing and Economics Reprints,* Am. Chem. Soc., San Francisco, CA, 1976, p. 212.

6. D. H. Lewis and D. R. Cowsar, "Principles of Controlled Release Pesticides, in H. B. Scher, Ed., *Controlled Release Pesticides (ACS Symp. Ser. 53),* Am. Chem. Soc., Washington, DC, 1977, p. 11.

7. G. O. Fanger, "General Background and History of Controlled Release," in N. F. Cardarelli, Ed., *Proc. Int. Controlled Release Pesticide Symp., University of Akron, Ohio,* 1974, p. 1.7.

8. R. W. Baker and H. K. Lonsdale, "Principles of Controlled Release," in F. W. Harris, Ed., *Proc. Int. Controlled Release Pesticide Symp. Wright State University, Dayton, OH,* 1975, p. 10.

9. A. M. Neogi and G. G. Allan, "Controlled-Release Pesticides: Concepts and Realization," *Advan. Exper. Med. Biol.* 47, 210 (1947).

10. G. Zweig, "Environmental Aspects of Controlled Release Pesticide Formulation, in H. B. Scher, Ed., *Controlled Release Pesticides (ACS Symp. Ser.* **53**), Am. Chem. Soc., Washington, DC, 1977, p. 45.

11. J. A. Bakan, "Microcapsule Drug Delivery Systems," in R. L. Kronenthal, Z. Oser, and E. Martin, Eds., *Polymers in Medicine and Surgery*, Plenum, New York, 1975, p. 213.

12. D. R. Paul, "Polymers in Controlled Release Technology," in D. R. Paul and F. W. Harris, Eds., *Controlled Release Polymeric Formulations (ACS Symp. Ser.* **33**), Am. Chem. Soc., Washington, DC, 1976, p. 9.

13. N. F. Cardarelli and S. V. Kanakkanatt, "Matrix Factors Affecting the Controlled Release of Pesticides from Elastomers," in H. B. Scher, Ed., *Controlled Release Pesticides (ACS Symp. Ser.* **53**). Am. Chem. Soc., Washington, DC, 1977, p. 64.

14. A. N. Neogi and A. G. Graham, "Controlled Release Pesticides: Concepts and Realization," in A. C. Tanquary and R. E. Lacey, Eds., *Controlled Release of Biologically Active Agents*, Plenum, New York, 1974, p. 210.

15. R. W. Baker and H. K. Lonsdale, "Membrane-Controlled Delivery Systems," in N. F. Cardarelli, Ed., *Proc. Int. Controlled Release Pesticide Symp., University of Akron, OH,* 1974. p. 40.1.

16. J. Crank and G. S. Park, Eds., *Diffusion in Polymers*, Academic, London, 1968.

17. J. Crank, *The Mathematics of Diffusion*, Oxford Univ. Press, London, 1956.

18. J. M. Folkman and D. M. Long, Jr., "The Use of Silicone Rubber as a Carrier for Prolonged Drug Therapy," *J. Surg. Res.*, **4**, 139 (1964).

19. J. Folkman and D. M. Long, Jr., "Drug Pacemakers in the Treatment of Heart Block," *Ann. N.Y. Acad. Sci.*, **111**, 857 (1964).

20. J. Folkman, D. M. Long, Jr., and R. Rosenbaum, "Silicone Rubber: A New Diffusion Property Useful for General Anesthesia," *Science*, **154**, 148 (1966).

21. J. Folkman, S. Winsey, and T. Moghul, "Anesthesia," *Science*, **154**, 148 (1966).

22. J. Folkman, V. H. Mark, F. Ervin, K. Suematsu, and R. Hagiwara, "Intravenous Gas Anesthesia by Diffusion through Silicone Rubber," *Anesthesiology*, **29**, 419 (1968).

23. E. M. Coutinho and A. R. da Silva, "One Year Contraception with Norgestrienone Subdermal Silastic Implants," *Fertil. Steril.*, **25**, 170 (1974).

24. S. Tejuja, U. Malhotra, and G. Bhinder, "A Preliminary Report on the Contraceptive Use of Subdermal Implants Containing Norethindrone," *Contraception*, **10**, 361 (1974).

25. E. M. Coutinho, A. R. da Silva, C. M. Carreira, M. C. Chaves, and J. Adeodato Filho, "Contraceptive Effectiveness of Silastic Implants Containing the Progestin R 2323," *Contraception*, **11**, 625 (1975).

26. E. Weiner and E. D. B. Johansson, "Plasma Levels of *d*-Norgestrel, Estradiol and Progesterone during Treatment with Silastic Implants Containing *d*-Norgestrel," *Contraception*, **14**, 81 (1976).

27. B. B. Pharriss, "Steroid Delivery Systems for Contraception," *J. Reprod. Med.*, **17**, 91 (1976).

28. L. L. Ewing, C. Desjardins, and L. G. Stratton, Testosterone Polydimethyl Siloxane Implants and Contraception in Male Rabbits," in J. Urquhart and F. E. Yates, Eds., *Temporal Aspects of Therapeutics*, Plenum, New York, 1973, p. 165.

29. L. L. Ewing, L. G. Stratton, and C. Desjardins, "Effect of Testosterone Polydimethyl-siloxane Implants upon Sperm Production, Libido and Accessory Sex Organ Function in Rabbits," *J. Reprod. Fertil.,* **35**, 245 (1973).

30. A. Zaffaroni, "Special Requirements for Hormone Releasing Intrauterine Devices," *Acta Endocrinol.,* **75**(Suppl. 185), 423 (1974).

31. H. A. Quigley, I. P. Pollack, and T. S. Harbin, "Pilocarpine Ocuserts, Long-Term Clinical Trials and Selected Pharmacodynamics," *Arch. Ophthalmol.,* **93**, 771 (1975).

32. D. J. Lyman and S. W. Kim, "Aqueous Diffusion through Partition Membranes," *J. Polym. Sci. Polym. Symp.,* **41**, 139 (1973).

33. M. A. Gonzales, J. Nematoblaki, W. L. Guess, and J. Autian, "Diffusion, Permeation, and Solubility of Selected Agents in and through Polyethylene," *J. Pharm. Sci.,* **56**, 1288 (1967).

34. D. R. Kalkwarf, M. R. Sikof, L. Smith, and N. Gordon, "Release of Progesterone From Polyethylene Devices *in vitro* and in Experimental Animals," *Contraception,* **6**, 424 (1972).

35. H. B. Hopfenberg, Ed., *Polymer Science and Technology,* Vol. 6, Plenum, New York, 1974.

36. F. Theeuwes, K. Ashida, and T. Higuchi, "Programmed Diffusional Release from Encapsulated Cosolvent System," *J. Pharm. Sci.,* **65**, 648 (1976).

37. M. D. de Gennaro, B. B. Thompson, and L. A. Luzzi, "Effect of Cross-linking Agents on the Release of Sodium Pantobarbital From Nylon Microcapsules," and references therein, in D. R. Paul and F. W. Harris, Eds., *Controlled Release Polymeric Formulations (ACS Symp. Ser.* **33**), Am. Chem. Soc., Washington, DC, 1976, p. 195.

38. L. A. Luzzi, M. A. Zoglio, and H. V. Moulding, "Preparation and Evaluation of the Prolonged Release Properties of Nylon Microcapsules," *J. Pharm. Sci.,* **59**, 338 (1970).

39. F. Bottari, G. di Colo, E. Nannipieri, M. F. Saettone, and M. F. Serafini, "Evaluation of a Dynamic Technique for Studying Drug-Macromolecule Interactions," *J. Pharm. Sci.,* **64**, 946 (1975).

40. D. R. Cowsar, O. R. Tarwater, and A. C. Tanquary, "Controlled Release of Fluoride for Dental Applications," in J. Andrade, Ed., *Hydrogels for Medical and Related Applications,* Am. Chem. Soc., Washington, DC, 1976, p. 180.

41. T. W. Brooks, E. Ashare, and D. W. Swenson, "Hollow Fibers as Controlled Release Vapor Devices," in J. C. Arthur, Ed., Textile and Paper Chemistry and Technology *(ACS Symp. Ser.* **49**), Am. Chem. Soc., Washington, DC, 1977, p. 111.

42. E. Ashare, T. W. Brooks, and D. W. Swenson, "Controlled Release from Hollow Fibers, in D. R. Paul and F. W. Harris, Eds., *Controlled Release Polymeric Formulations (ACS Symp. Ser.* **33**), Am. Chem. Soc., Washington, DC, 1976, p. 273.

43. M. J. Coplan and T. W. Brooks, U.S. Patent 4,017,030 (1977).

44. A. S. Michaels, "Therapeutic Systems for Controlled Administration of Drugs: A New Application of Membrane Science," in H. P. Hopfenberg, Ed., *Permeability of Plastic Films and Coatings,* Plenum, New York, 1975.

45. S. K. Chandrasekaran and J. Shaw, "Design of Transdermal Therapeutic Systems," in E. M. Pearch and J. R. Schaefgen, Eds., *Contemporary Topics in Polymer Science,* Vol. 2, Plenum, New York, 1977.

46. M. K. Addapeddi et al., "Cervical Hydrogel Dilator: A New Delivery System for Prostaglandins", in A. S. Tanquary and R. E. Lacey, Eds., *Controlled Release of Biologically Active Agents*, Plenum, New York, 1974, p. 165.

47. Anonymous, "Slow Release Achieved for Macromolecules, " *C & E News*, September 5, 1977, p. 26.

48. M. Tollar et al., *J. Biomed. Mater. Res.*, **3**, 305 (1969).

49. J. Drobnik et al., *J. Biomed. Mater. Res.*, **8**, 45 (1974).

50. R. S. Molday et al., *Nature*, **249**, 81 (1974).

51. A. S. Obermayer and L. D. Nicholas, "Controlled Release from Ultramicroporous Cellulose Triacetate," in D. R. Paul and F. W. Harris, Eds., *Controlled Release Polymeric Formulations (ACS Symp. Ser.* **33**), Am. Chem. Soc., Washington, DC, 1976, p. 303.

52. L. D. Nichols, "Poroplastic and Sustrelle: Controlled Release Vehicles Having Broad Compatibility with Dissolved and Precipitated Pesticides," in N. F. Cardarelli, Ed., Proc. Int. Controlled Release Pesticide Symp., *University of Akron, OH, 1974.*

53. T. W. Brooks, "Controlled Vapor Release from Hollow Fibers: Theory and Applications with Insect Pheromones," in A. F. Kydonieus, Ed., *Controlled Release Technologies: Methods, Theory, and Applications*, CRC Press, Boca Raton, FL, 1980, p. 165.

54. A. S. Obermayer and L. D. Nichols, "Controlled Release from Ultramicroporous Cellulose Triacetate," in D. R. Paul and F. W. Harris, Eds., *Controlled Release Polymeric Formulations (ACS Sym. Ser.* **33**), Am. Chem. Soc. Washington, DC, 1976, p. 303.

55. L. D. Nichols, "Poroplastic® and Sustrelle®, Controlled Release Vehicles Having Broad Compatibility with Dissolved and Precipitated Pesticides," in N. F. Cardarelli, Eds., *Proc. Int. Controlled Release Pesticide Symp. University of Akron, OH*, 1974.

56. A. S. Obermayer and L. D. Nichols, "Controlled Release from Poroplastic®, Sustrelle®, and LiquiPowder®," Society of Cosmetic Chemists Annual Meeting, New York, 1976.

57. A. S. Obermayer, "Controlled Release from Ultramicroporous Triacetate," in A. F. Kydonieus, Ed., *Controlled Release Technologies: Methods, Theory, and Applications*, CRC Press, Boca Raton, FL, 1980 p. 239.

58. B. D. Ratner and A. S. Hoffman, "Synthetic Hydrogels for Biomedical Applications," in J. Andrade, Ed., *Hydrogels for Medical and Related Applications*, Am. Chem. Soc., Washington, DC, 1976, p. 1.

59. H. Yasuda, C. E. Lamaze, and A. Peterlin, Diffusive and Hydraulic Permeabilities of Water in Water-Swollen Polymer Membranes," *J. Polym. Sci. Part A*, **9**, 1117 (1971).

60. R. Y. S. Chen, "Diffusion Coefficients and Swelling Behavior of Cross-linked Poly(2-hydroxyethyl methacrylamide)," *Polym. Prepr. Am. Chem. Soc. Div. Polym. Chem.*, **15**(2), 387 (1974).

61. B. D. Ratner and I. F. Miller, "Transport through Crosslinked Poly(2-hydroxyethyl methacrylate) Hydrogel Membranes," *J. Biomed. Mater. Res.*, **7**, 353 (1973).

62. S. J. Wisniewski, D. E. Gregonis, S. W. Kim, and J. D. Andrade, "Diffusion through Hydrogel Membranes. I. Permeation of Water through Poly(2-hydroxy ethyl methacrylate) and Related Polymers", in J. D. Andrade, Ed., *Hydrogels for Medical and Related Applications*, Am. Chem. Soc., Washington, DC, 1976.

63. B. D. Halpern, O. Solomon, L. Kopec, E. Korostoff, and J. L. Ackerman, "Release of Inorganic Fluoride Ion from Rigid Polymer Matrices," in D. R. Paul and F. W. Harris, Eds., *Controlled Release Polymeric Formulations*, Am. Chem. Soc., Washington, DC, 1976.

64. J. Drobnik, P. Spacek, and O. Wichterle, "Diffusion of Antitumor Drugs through Membranes from Hydrophilic Methacrylate Gels," *J. Biomed. Mater. Res.*, 8, 45 (1974).

65. B. S. Levowitz, J. N. Laguere, W. S. Calem, F. E. Gould, J. Scherrer, and H. Schoenfeld, "Biologic Compatibility and Applications of Hydron," *Trans. Am. Soc. Artif. Intern. Organs*, 14, 82 (1968).

66. S. M. Lazarus, J. N. Laguerre, H. Kay, S. Weinberg, and B. S. Levowitz, "A Hydrophilic Polymer-Coated Antimicrobial Urethral Catheter," *J. Biomed. Mater. Res.*, 5, 129 (1971).

67. M. Tollar, M. Stol, and K. Klimet, "Surgical Suture Coated with a Layer of Hydrophilic Hydron Gel," *J. Biomed. Mater. Res.*, 3, 305 (1969).

68. V. Majku, F. Horakora, F. Vymola, and M. Stol, "Employment of Hydron Polymer Antibiotic Vehicle in Otolaryngology," *J. Biomed. Mater. Res.*, 3, 443 (1969).

69. H. E. Kaufman, M. H. Votila, A. R. Casset, T. O. Wood, and E. D. Varnell, "Medical Uses of Soft Contact Lenses," in A. R. Gasset and H. E. Kaufman, Eds., *Soft Contact Lenses*, C.V. Mosby, S. Louis, 1972, Chap. 22.

70. S. M. Podos, B. Becker, C. Asseff, and J. Hartstein, "Pilocarpine Therapy with Soft Contact Lenses," *Am. J. Ophthalmol.*, 73, 336 (1972).

71. K. D. Colter, A. T. Bell, and M. Shen, "Control of the Pilocarpine Release Rate through Hydrogels by Plasma Treatment," *Biomater. Med. Devices Artif. Organs*, 5, (1977).

72. H. Scott, P. L. Kronick, R. C. May, R. H. Davis, and H. Balin, "Construction and Properties of Hydrogel-Graft-Coated Copper-Bearing Intrauterine Devices for Rabbits," *Biomater. Med. Devices Artif. Organs*, 1, 681 (1973).

73. I. C. Craig and H. Chen, "On a Theory for the Passive Transport of Solute through Semipermeable Membranes," *Proc. Nat. Acad. Sci. USA*, 67, 702 (1972).

74. P. C. Farrell and A. L. Babb, "Estimation of the Permeability of Cellulosic Membranes from Solute Dimension and Diffusivities," *J. Biomed. Mater. Res.*, 7, 275 (1973).

75. W. M. Muir, R. A. Gray, J. M. Courtney, and P. D. Ritchie, "Permselective Dialysis Membranes. II. Films Based on Acrylic Acid–*n*-Butyl Methacrylate Copolymers—A Critical Comparison with Cellulosic Films," *J. Biomed. Mater. Res.*, 7, 3 (1973).

76. D. J. Lyman and S. W. Kim, "Aqueous Diffusion through Partition Membranes," *J. Polym. Sci. Polym. Symp.*, 41, 139 (1973).

77. L. C. Craig and W. Konigsberg, "Dialysis Studies. III. Modification of Pore Size and Shape in Cellophane Membranes," *J. Phys. Chem.*, 65, 166 (1961).

78. W. G. Bradley and G. L. Wilkes, "Some Mechanical Property Considerations of Reconstituted Collagen for Drug Release Supports," *Biomater. Med. Devices Artif. Organs*, 5, 159 (1977).

79. P. Meares, Ed., *Membrane Separation Processes*, Elsevier, Amsterdam, 1976.

80. A. S. Michaels, "Therapeutic Systems for Controlled Administration of Drugs: A New Application of Membrane Science," in H. P. Hopfenberg, Ed., *Permeability of Plastic Films and Coatings*, Plenum, New York, 1975, p. 409.

81. W. P. O'Neill, "Membrane Systems," in A. F. Kydonieus, Ed., *Controlled Release Technologies: Methods, Theory, and Applications*, Vol. I, CRC Press, Boca Raton, FL, 1980, p. 129.

82. F. W. Harris, "Preparation of Plastic, Controlled Release Pesticide Formulations," in N. F. Cardarelli, Ed., *Proc. Int. Controlled Release Pesticide Symp.*, Univ. Akron, Akron, OH, 1976, p. 1.33.

83. N. F. Cardarelli, "Compounding Methods for Controlled Release Elastomers," in N. F. Cardarelli, Ed., *Proc. Int. Controlled Release Pesticide Symp.*, Univ. Akron, Akron, OH, 1976, p. 1.44.

84. S. Yolles et al., in A. C. Tanquary and R. E. Lacey, Eds., *Controlled Release of Biologically Active Agents*, Plenum, New York, 1974.

85. G. G. Allan et al., "Polymeric Drugs for Plants," in L. G. Donarums and O. Vogi, Eds., *Polymeric Drugs*, Plenum, New York, 1974.

86. G. G. Allan, J. W. Beer, and M. J. Cousin, "Controlled Release of Herbicides from Biodegradable Substrates," in H. E. Scher, Ed., *Controlled Release Pesticides (ACS Symp. Ser. 53)*, Am. Chem. Soc., Washington, DC, 1977, p. 94.

87. B. S. Shasha, W. M. Doane, and C. R. Russell, "Starch-Encapsulated Pesticides for Slow Release," *J. Polym. Sci.*, 14, 417 (1976).

88. W. M. Doane, B. S. Shasha, and C. R. Russell, "Encapsulation of Pesticides with a Starch Matrix," in H. B. Scher, Ed., *Controlled Release Pesticides (ACS Symp. Ser. 53)*, Am. Chem. Soc., Washington, DC, 1977.

89. H. T. DelliColli, "Controlled Release of Pesticides from Kraft Lignin Carriers," in H. B. Scher, Ed., *Controlled Release Pesticides (ACS Symp. Ser. 53)*, Am. Chem. Soc., Washington, DC, 1977.

90. H. B. Hopfenberg, "Controlled Release from Erodible Slabs, Cylinders and Spheres," in D. R. Paul and F. W. Harris, Eds., *Controlled Release Polymeric Formulations (ACS Symp. Ser. 33)*, Am. Chem. Soc., Washington, DC, 1976, p. 26.

91. J. H. R. Woodland, S. Yolles, D. Blake, M. Helrich, and F. J. Meyer, "Long-Acting Delivery Systems for Narcotic Antagonists," *J. Med. Chem.*, 16, 899 (1973).

92. A. D. Schwope, D. L. Wise, and J. F. Howes, "Lactic/Glycolic Acid Polymers as Narcotic Antagonist Delivery Systems," *Life Sci.*, 17, 1878 (1975).

93. R. G. Sinclair and G. M. Gynn, "Preparation and Evaluation of Glycolic and Lactic Acid-Based Polymers," *Final Sci. Rep.*, Part II, Contract No. NADA 17-72-2066, U.S. Army Inst. Dental Res., Walter Reed Army Med. Center, Washington, DC, 1972, p. 6.

94. S. Yolles, "Time Release Depot for Anticancer Drugs: Release of Drugs Covalently Bonded to Polymers," *J. Parenter. Drug Assoc.*, 32, 188 (1978).

95. A. Schindler, J. F. Coat, G. Kimble, C. G. Pitt, M. E. Wall, and R. Zweidinger, "Biodegradable Polymers for Sustained Drug Delivery," in E. M. Pearce and J. R. Schaefgen, Eds., *Contemporary Topics in Polymer Science*, Vol. 2, Plenum, New York, 1977, p. 264.

96. R. K. Kulkarni, K. Pani, C. Neuman, and F. Leonard, "Poly(lactic acid) for Surgical Implants," *Tech. Repr.* No. 6608, Walter Reed Army Med. Center, Washington, DC, 1966.

97. W. B. Ramsey and D. F. DeLapp, U. S. Patent 3,781,349 (1973).

98. S. Yolles, U. S. Patent 3,887,699 (1975).

99. T. M. Jackanicz, H. H. Nash, D. L. Wise, and J. B. Gregory, "Poly(lactic acid) as a Biodegradable Carrier for Contraceptive Steriods," *Contraception*, 8, 227 (1973).

100. S. Yolles, T. D. Leafe, and J. Meyer, "Controlled Release of Anticancer Agents," *J. Pharmacol. Sci.,* **64**, 115 (1975).

101. A. K. Schneider, Fr. Patent 1,478,694 (1967).

102. T. D. Leafe, S. Sarner, J. H. R. Woodland, S. Yolles, D. H. Blake, and F. J. Meyer, "Narcotic antagonists," *Advan. Biochem. Psychopharm.* **8**, 569 (1973).

103. S. Yolles, T. D. Leafe, J. H. R. Woodland, and F. J. Meyer, "Release Rates of Naltrexone from PLA," *J. Pharmacol. Sci.,* **64**, 348 (1975).

104. S. Yolles, T. D. Leafe, M. S. Sartori, M. Torkelson, and L. Ward, "Controlled Release of Biologically Active Agents," in D. R. Paul and F. W. Harris, Eds., *Controlled Release Polymeric Formulations (ACS Symp. Ser.* **33**), Am. Chem. Soc., Washington, DC, 1976, p. 125.

105. R. G. Sinclair, "Slow Release Pesticide System," *Environ. Sci. Technol,* **7**, 955 (1973).

106. F. P. Furgivele, "Ophthalmic Use of 'Dexon'," *Ann Ophthalmol.,* **6**, 1219 (1974).

107. T. P. Kuo, "Clinical Use of PGA in Vascular Surgery," *J. Formosan Med. Assoc.,* **73**, 45 (1974).

108. E. E. Schmitt, U.S. Patents 3,464,158 (1969); 3,739,773 (1973).

109. E. E. Schmitt and R. A. Polistina, U.S. Patent 3,991,766 (1976).

110. E. E. Schmitt and R. A. Polistina, U.S. Patent 3,297,033 (1967).

111. E. E. Schmitt, S. Afr. Patent 71-08.150 (1972).

112. A. Glick, U.S. Patent 3,772,420 (1973).

113. D. Wasserman, U.S. Patent 3,839,297 (1974).

114. R. G. Sinclair and G. M. Gynn, *U.S. Army Med. Res. Develop. Command Rep.,* Rm 8H089, Forrestal Building, Washington, DC, 1972.

115. A. F. Hegyeli, "Biodegradation of PLA/PGA Implant Materials," *J. Biomed. Mater. Res.,* **7**, 205 (1973).

116. D. E. Cutright, "Degradation of Polymers and Copolymers of PLA/PGA," *Oral Surg.,* **37**, 142 (1974).

117. R. B. Sinclair, "Slow Release Pesticide System," *Environ. Sci. Technol.,* **7**, 956 (1973).

118. T. J. Roseman, "Release of Steroids from a Silicone Polymer," *J. Pharmacol. Sci.,* **61**, 46 (1972).

119. J. Haleblian, R. Runkel, N. Mueller, J. Christopherson, and K. Ng, "Steroid Release from Silicone Elastomer Containing Excess Drug in Suspension," *J. Pharmacol. Sci.,* **60**, 541 (1971).

120. Y. W. Chien, H. J. Lambert, and D. E. Grant, "Controlled Drug Release from Polymeric Devices. I. Technique for Rapid *in vitro* Release Studies," *J. Pharmacol. Sci.,* **63**, 365 (1974).

121. V. W. Winkler, S. Borodkin, S. K. Webel, and J. T. Mannebach, "*In vitro* and *in vivo* Considerations of a Novel Matrix-Controlled Bovine Progesterone-Releasing Intra-Vaginal Device," *J. Pharmacol. Sci.,* **66**, 816 (1977).

122. R. W. Baker and H. K. Lonsdale, "Membrane Controlled Delivery Systems," in N. F. Cardarelli, Ed., *Proc. Int. Controlled Release Pesticide Symp. Univ. Akron, Ohio,* 1974.

123. S. J. Desai, A. P. Simonelli, and W. I. Higuchi, "Investigation of Factors Influencing Release of Solid Drug Dispersed in Inert Matrices," *J. Pharmacol. Sci.,* **54**, 1459 (1965).

124. S. J. Desai, P. Singh, A. P. Simonelli, and W. I. Higuchi, "Investigation of Factors Influencing Release of Solid Drug Dispersed in Inert Matrices. III. Quantitative Studies Involving the Polyethylene Plastic Matrix," *J. Pharmacol. Sci.,* 55, 1230 (1966).

125. S. J. Desai, P. Singh, A. P. Simonelli, and W. I. Higuchi, "Investigation of Factors Influencing Release of Solid Drug Dispersed in Inert Matrices. IV. Some Studies Involving the Polyvinyl Chloride Matrix," *J. Pharmacol. Sci.,* 55, 1235 (1966).

126. J. B. Schwartz, A. P. Simonelli, and W. I. Higuchi, "Drug Release from Wax Matrices. I. Analysis of Data with First-Order Kinetics and with the Diffusion Controlled Model," *J. Pharmacol. Sci.,* 57, 274 (1968).

127. J. B. Schwartz, A. P. Simonelli, and W. I. Higuchi, "Drug Release from Wax Matrices. II. Application of a Mixture Theory to the Sulfanilamide-Wax System," *J. Pharmacol. Sci.,* 57, 278 (1968).

128. B. Farhadieh, S. Borodkin, and J. D. Buddenhagen, "Drug Release from Methyl Acrylate-Methyl Methacrylate Copolymer Matrix. I. Kinetics of Release," *J. Pharmacol. Sci.,* 60, 209 (1971).

129. H. Lapidus and N. G. Lordi, "Some Factors Affecting the Release of a Water Soluble Drug from a Compressed Hydrophilic Matrix," *J. Pharmacol. Sci.,* 55, 840 (1966).

130. H. Lapidus and N. G. Lordi, "Drug Release from Compressed Hydrophilic Matrices," *J. Pharmacol. Sci.,* 57, 1292 (1968).

131. S. J. Desai, A. P. Simonelli, and W. I. Higuchi, "Investigation of Factors Influencing Release of Solid Drug Dispersed in Inert Matrices," *J. Pharmacol. Sci.,* 54, 1459 (1965).

132. A. S. Michaels and H. J. Bixler, "Membrane Permeation: Theory and Practice," in E. S. Perry, Ed., *Progress in Separation and Purification*, Vol. 1, Interscience, New York, 1968, p. 143.

133. S. B. Tuwiner, *Diffusion and Membrane Technology*, Van Nostrand-Reinhold, New York, 1962.

134. J. Crank and G. S. Park, Eds., *Diffusion in Polymers*, Academic, New York, 1968.

135. A. Leo, C. Hansch, and D. Elkins, "Partition Coefficients and their Uses," *Chem. Rev.,* 71, 525 (1971).

136. T. J. Roseman and S. H. Yalkowsky, "Importance of Solute Partitioning on the Kinetics of Drug Release from Matrix Systems," in D. R. Paul and F. W. Harris, Eds., *Controlled Release Polymeric Formulations (ACS Symp. Ser. 33)*, Am. Chem. Soc., Washington, DC, 1976, p. 33.

137. Y. W. Chien, "Thermodynamics of Controlled Drug Release from Polymeric Delivery Devices," in D. R. Paul and F. W. Harris, Eds., *Controlled Release Polymeric Formulations (ACS Symp. Ser. 33)*, Am. Chem. Soc., Washington, DC, 1976, p. 53.

138. N. F. Cardarelli and N. F. Neff, "Biocidal Elastomeric Compositions," U.S. Patent 3,639,583 (1972).

139. N. F. Cardarelli, "Slow Release Molluscicides and Related Materials," in T. C. Cheng, Ed., *Molluscicides in Schistosomiasis Control*, Academic, New York, 1974, p. 177.

140. N. F. Cardarelli, U. S. Patent 3,851,053 (1974).

141. W. E. Thompson, "Field Tests of Slow Release Herbicides," in N. F. Cardarelli, Ed., *Proc. Int. Controlled Release Pesticide Symp., Univ. Akron, OH*, 1974, Rep. 15.

141a. K. G. Das, *Proc. Int. Controlled Release Pesticide Symp.*, Nat. Bur. Std., Maryland, 1978, p. 3.29.

142. C. G. L. Furmidge, A. C. Hill, and J. M. Osgerby, "Physicochemical Aspects of the Availability of Pesticides in the Soil. I, Leaching of Pesticides from Granular Formulations," *J. Sci. Food Agric.,* 17, 518 (1966).

143. B. D. Dinman, "'Non-concept' of 'no-threshold': Chemicals in the Environment," *Science,* 175, 495 (1972).

144. H. F. Kraybill, "Pesticide Toxicity and Potential for Cancer: A Proper Perspective," *Pest Control,* 43(12), 1975).

145. N. F. Cardarelli, "Hypothesis Concerning Chronic Intoxication," in N. F. Cardarelli, Ed., *Proc. Int. Controlled Release Pesticide Symp., Wright State Univ., Dayton, OH,* 1975, p. 349.

146. J. Crank and G. S. Park, "Methods of Measurement," in J. Crank and G. S. Park, Eds., *Diffusion of Polymers,* Academic, New York, 1968, p. 1.

147. R. W. Richards, "The Permeability of Polymers to Gases, Vapours, and Liquids," *Explosives Res. Develop. Establ. (Brit. Ministry of Defense) Tech. Rep.* No. 135, NTIS AD-767 627, March 1973.

148. R. W. Baker, H. K. Lonsdale, and R. M. Gale, "Membrane-Controlled Delivery Systems," in N. F. Cardarelli, Ed., *Proc. Int. Controlled Release Pesticide Symp., Univ. Akron, Akron, OH,* 1974, p. 40.1.

149. R. W. Baker and H. K. Lonsdale, "Controlled Release: Mechanisms and Rates," in A. C. Tanquary and R. E. Lacey, Eds., *Controlled Release of Biologically Active Agents,* Plenum, New York, 1974, ch. 2.

150. F. W. Harris, "Theoretical Aspects of Controlled Release," in N. F. Cardarelli, Ed., *Proc. Int. Controlled Release Pesticide Symp., Univ. Akron, Akron, OH,* 1974, p. 8.1.

151. J. Crank, *The Mathematics of Diffusion,* Oxford Univ. Press, London, 1956.

152. R. M. Barrer, "Diffusion and Permeation in Heterogeneous Media," in J. Crank and G. S. Park, Eds., *Diffusion of Polymers,* Academic, New York, 1968, p. 165.

153. A. F. Kydonieus, "The Effect of Some Variables on the Controlled Release of Chemicals from Polymeric Membranes," in H. B. Scher, Ed., *Controlled Release Pesticides (ACS Symp. Ser.* 53), Am. Chem. Soc., Washington, DC, 1977, p. 152.

154. A. F. Kydonieus, A. R. Quisumbing, and S. Hyman, "Application of a New Controlled Release Concept in Household Products," in D. R. Paul and F. W. Harris, Eds., *Controlled Release Polymeric Formulations (ACS Symp. Ser.* 33), Am. Chem. Soc., Washington, DC, 1976, p. 295.

155. A. F. Kydonieus, A. Rofheart, and S. Hyman, "Marketing and Economic Considerations for Hercon Consumer and Industrial Controlled Release Products," in *Chemical Marketing and Economic Preprints of Symposium on Economics and Market Opportunities for Controlled Release Products,* ACS Chem. Marketing and Economics Div., San Francisco, 1976, p. 140.

156. M. Beroza and N. Green, "Synthetic Chemicals as Insect Attractants," *Advan. Chem. Ser.,* 41, 11 (1963).

157. M. Beroza and E. F. Knipling, "Gypsy Moth Control with the Sex Attractant Pheromone." *Science,* 177, 19 (1972).

158. M. S. Mayer and J. R. McLaughlin, *An Annotated Compendium of Insect Sex Pheromones (Fla. Agric. Exp. Sta. Monograph Ser.* 6), Univ. Florida, Gainesville, FL, 1975.

159. F. J. Mulhern, Administrator, Trial Boll Weevil Eradication Program Environmental Statement, USDA, APHIS, (ADM)-71-1, Washington, DC, 1975.

160. W. P. Scott, E. P. Lloyd, J. O. Bryson, and T. B. Davich, "Trap Crops for the Suppression of Low Density Overwintered Boll Weevil Populations," *J. Econ. Entomol.*, 67, 281 (1974).

161. D. D. Hardee, G. H. McKibben, and P. M. Huddleston, "Grandlure for Boll Weevils: Controlled Release with a Laminated Plastic Dispenser," *J. Econ. Entomol.*, 68, 477 (1975).

162. W. L. Johnson, G. H. McKibben, J. Rodriguez, and T. Davich, "Boll Weevil: Increased Longevity of Grandlure Using Different Formulations and Dispensers," *J. Econ. Entomol.*, 69 263 (1976).

163. T. B. Davich and G. H. McKibben, "Belt-Wide Grandlure Formulation Test," presented at the Entomological Soc. Am. Annual Meeting, New Orleans, 1975.

164. F. W. Michelotti, A New Japanese Beetle Trap Containing Pheromone and Floral Lure as Synergistic Attractants," presented at the Entomological Soc. Am. Annual Meeting, Denver, 1979.

165. G. N. Lanier, "Protection of Elm Groves by Surrounding Them with Multilure Baited Sticky Traps," *Bull. Entomol. Soc. Amer.*, 25, 109 (1979).

167. R. A. Cuthbert and J. W. Peacock, "The Forest Service Program for Mass-Trapping *Scolytus multistriatus,*" *Bull. Entomol. Soc. Am.*, 25, 109 (1979).

168. A. Bakke and L. Riege, "Mass Trapping of the Spruce Bark Beetle. *Ips typographus*, in Norway as Part of an Integrated Control Program." in preparation.

169. A. Bakke, "The Pheromone of the Spruce Bark Beetle, *Ips typographus*, and Its Potential Use in the Suppression of Beetle Populations," in preparation.

170. A. R. Quisumbing, A. F. Kydonieus, D. R. Calsetta, and J. B. Haus, "Hercon Lure 'N Kill Flytape: A Non-fumigant Insecticidal Strip Containing Attractants," in N. F. Cardarelli, Ed., *Proc. Int. Controlled Release Pesticides Symp., Univ. Akron, OH*, 1976, p. 3.40.

171. A. F. Kydonieus and A. R. Quisumbing, "Multilayered Laminated Structures," in A. A. F. Kydonieus, Ed., *Controlled Release Technologies: Methods, Theory and Applications*, Vol. 1, CRC Press, Boca Raton, FL, 1980, p. 183.

172. S. Nakagawa, E. J. Harris, and T. Urago, "Controlled Release of Trimedlure from a Three-Layer Laminated Plastic Dispenser," *J. Econ. Entomol.*, 72, 625 (1979).

173. K. W. Vick, J. A. Coffelt, and M. A. Sullivan, "Disruption of Pheromone Communication in the *Angoumis* Grain Moth with Synthetic Female Sex Pheromone," *Environ. Entomol*, 7, 528 (1978).

174. T. J. Henneberry, J. M. Gillespie, L. A. Bariola, H. M. Flint, G. D. Butler, Jr., P. D. Lingren, and A. F. Kydonieus, "Mating Disruption as a Method of Controlling Pink Bollworm and Tobacco Budworm on Cotton," in preparation.

175. T. J. Henneberry, J. M. Gillespie, L. A. Bariola, H. M. Flint, G. D. Butler, Jr., P. D. Lingren, and A. F. Kydonieus, "Gossyplure in Laminated Plastic Formulations for Mating Disruption and Pink Bollworm Control," *J. Econ. Entomol.* forthcoming.

176. T. J. Henneberry, J. M. Gillespie, L. A. Bariola, H. M. Flint, G. D. Butler, Jr., P. D. Lingren, and A. F. Kydonieus, "Recent Progress with Gossyplure for Pink Bollworm Mating Disruption," in *Proc. 8th Desert Cotton Insect Symp.*, El Centro, CA, 1980.

177. G. E. Daterman and L. L. Sower, "Douglas Fir Tussock Moth Pheromone Research

Using Controlled Release System," in R. Goulding, Ed., *Proc. Intern. Controlled Release Pesticide Symp., Oregon State University, Corvallis*, 1977, p. 68.

178. M. Beroza, E. C. Paszek, E. R. Mitchell, B. A. Bierl, J. R. McLaughlin, and D. L. Chambers, "Tests of a Three-Layer Laminated Plastic Bait Dispenser for Controlled Emission of Attractants from Insect Traps," *Environ. Entomol.*, 3, 926 (1974).

179. M. Beroza, C. S. Hood, D. Trefrey, D. E. Leonard, E. F. Knipling, and W. Klassen, "Field Trials with Disparlure in Massachusetts to Suppress Mating of the Gypsy Moth," *Environ. Entomol.*, 4, 705 (1975).

180. M. Beroza, E. C. Paszek, D. DeVilbiss, B. A. Bierl, and J. G. R. Tardif, "A Three-Layer Laminated Plastic Dispenser of Disparlure for Use in Traps for Gypsy Moths, *Environ. Entomol.*, 4, 712 (1975).

181. J. R. Plimmer, C. P. Schwalbe, E. C. Paszek, B. A. Bierl, R. E. Webb, S. Marumo, and S. Iwaki, "Contrasting Effectiveness of (+) and (±) Enantiomers of Disparlure for Trapping Native Populations of Gypsy Moth in Massachusetts, *Environ. Entomol.*, 7, 815 (1978).

182. B. A. Bierl, E. D. DeVilbiss, and J. R. Plimmer, "Use of Pheromones in Insect Control Programs: Slow Release Formulations," in D. R. Paul and F. W. Harris, Eds., *Controlled Release Polymeric Formulations (ACS Symp. Ser. 33)*, Am. Chem. Soc., Washington, DC, 1976, p. 265.

183. J. R. Plimmer, B. A. Bierl, and C. P. Schwalbe, "Controlled Release of Pheromone in the Gypsy Moth Program," in H. B. Scher, Ed., *Controlled Release of Pesticides (ACS Symp. Ser. 53)*, Am. Chem. Soc., Washington, DC, 1977. p. 168.

184. R. E. Webb, C. W. Dull, C. W. McComb, B. A. Bierl, and J. R. Plimmer, "Gypsy Moth Mating Suppressed by Disparlure Emitted from Laminated Plastic Dispenser," unpublished.

185. R. E. Webb, C. P. Schwalbe, B. B. Leonhardt, and W. McLane, "Aerial Application of Disparlure Dispensers," unpublished.

186. S. L. Clement, B. S. Schmidt, and E. Levine, "Monitoring the Flight Activities of the Black Cutworm with Blacklight and Pheromone Traps," presented at the Entomol. Soc. Am. Annual Meeting, Denver, 1979.

187. F. C. Tingle and E. R. Mitchell, "Controlled Release Plastic Strips Containing (Z)-9-Dodecen-1-ol Acetate for Attracting *Spodoptera Frugiperda*," *J. Chem. Ecol.*, 4; 41 (1978).

188. D. E. Hendricks, A. W. Hartstack, and J. R. Raulston, "Compatibility of Virelure and Looplure Dispensed from Traps for Cabbage Looper and Tobacco Budworm Survey," *Environ. Entomol.*, 6, 566 (1977).

189. J. D. Lopez, Jr., T. N. Shaver, and A. W. Hartstack, Jr., "Evaluation of Dispensers for the Pheromone of *Heliothis zea*," in press.

190. A. W. Hartstack, Jr., Lopez, J. D., J. A. Klun, J. Witz, T. N. Shaver, and J. R. Plimmer, "New Trap Designs and Pheromone Bait Formulations for *Heliothis*," *Proc. Beltwide Cotton Prod. Res. Conf.*, 1980, p. 132.

191. A. F. Kydonieus, B. A. Bierl-Leonhardt, J. R. Plimmer, M. W. Barry, and A. R. Quisumbing, "Cotton Insects: Dispenser Development and Disruption of Mating Trials," in R. Baker, Ed., *Proc. 6th Int. Symp. Controlled Release Bioactive Materials, New Orleans, Louisiana*, 1980, p. IV-13.

192. D. E. Hendricks, A. W. Hartstack, and T. N. Shaver, "Effect of Formulations and Dis-

pensers on Attractiveness of Virelure to the Tobacco Budworm," *J. Chem. Ecol.,* **3,** 497 (1977).

193. J. P. Hollingsworth, A. W. Hartstack, D. R. Buck, and D. E. Hendricks, "Electric and Non-Electric Moth Traps Baited with the Synthetic Sex Pheromone of the Tobacco Budworm," U.S. Dept. Agric. Pub. ARS-S-173, Feb. 1978.

194. D. E. Hendricks, A. W. Hartstack, and T. N. Shaver, "Effect of Formulations and Dispensers on Attractiveness of Virelure to the Tobacco Budworm," *J. Chem. Ecol.,* **3,** 497 (1977).

195. C. R. Gentry, B. A. Bierl, and J. L. Blythe, "Air Permeation Field Trials with the Oriental Fruit Moth Pheromone," in N. F. Cardarelli, Ed., *Proc. Int. Controlled Release Pesticide Symp., Univ. Akron, OH,* 1976, p. 3.22.

196. C. R. Gentry, C. E. Yonce, and B. A. Bierl-Leonhardt, "Oriental Fruit Moth: Mating Disruption Trials with Pheromone," manuscript submitted for publication.

197. G. E. Daterman, "Control of Western Pineshoot Borer Damage by Mating Disruption—A Reality," manscript submitted for publication.

198. P. J. Landolt, C. E. Curtis, J. A. Coffelt, K. W. Vick, and R. Doolittle, "Disruption of Sexual Communication in the Navel Orangeworm, *Amyelois transitella,* in Almond Orchards," presented at the Entomol. Soc. Am. Annual Meeting, Denver, 1979.

199. D. T. Johnson, "Development of Control Strategies for Grape Root Borer," presented at the Entomological Soc. Amer. Southeastern Branch Meeting, Biloxi, MI, 1980.

200. C. E. Yonce and C. R. Gentry, "Disruption of Mating of Peachtree Borer," manuscript submitted for publication.

201. C. R. Gentry, B. A. Bierl-Leonhardt, and J. R. McLaughlin, "Trapping and Air Permeation with Pheromone to Combat Peach Orchard Pests," presented at the Entomol. Soc. Am. Southeastern Branch Meeting, Biloxi, MI, 1980.

202. C. R. Gentry, B. A. Bierl-Leonhardt, J. L. Blythe, J. R. McLaughlin, and J. R. Plimmer, "Air Permeation Tests with (Z,Z)-3,13-Octadecadien-1-ol Acetate for Reduction in Trap Catch of Peachtree and Lesser Peachtree Borer Moths," manuscript in preparation.

203. H. Arn, A. Schmid, C. Jaccard, B. A. Bierl-Leonhardt, and S. Rauscher, "Mating Reduction in Free-Living Grape Moths (*Eupoecilia ambiguella* Hb.) by Disruption of Pheromone Communication with (Z)-9-Dodecenyl Acetate," *Compte-rend. Reunion pheromones sexuelles des insectes et mediateurs chimiques,* INRA, Antibes, 1978.

204. M. Beroza, E. C. Paszek, D. DeVilbiss, B. A. Bierl, and J. G. R. Tardif, "A Three-Layer Laminated Plastic Dispenser of Disparlure for Use in Traps for Gypsy Moths," *Environ. Entomol.,* **4,** 712 (1975).

205. B. A. Bierl, M. Beroza, and C. W. Collier, "Potent Sex Attractant of the Gypsy Moth, *Porthetria dispar* (L.): Its Isolation, Identification and Synthesis," *Science,* **170,** 87 (1970).

206. M. Beroza, C. S. Hood, D. Trefrey, D. E. Leonard, E. F. Knipling, and W. Klassen, "Field Trials with Disparlure in Massachusetts to Suppress Mating of the Gypsy Moth" *Environ. Entomol.,* **4,** 705 (1975).

207. C. P. Schwalbe, E. A. Cameron, D. J. Hall, J. V. Richerson, M. Beroza, and L. J. Stevens, "Field Tests of Microencapsulated Disparlure for Suppression of Mating among Wild and Laboratory-Reared Gypsy Moths," *Environ. Entomol.,* **3,** 589 (1974).

208. R. E. Webb, C. W. Dull, C. W. McComb, B. A. Bierl, and J. R. Plimmer, "Gypsy Moth Mating Suppressed by Disparlure Emitted from Laminated Plastic Dispenser," 1978, submitted for publication.

209. B. A. Bierl, E. D. DeVilbiss, and J. R. Plimmer, "Use of Pheromones in Insect Control Programs: Slow Release Formulations," in D. R. Paul and F. W. Harris, eds., *Controlled Release Polymeric Formulations (ACS Symp. Ser.* **33**), Am. Chem. Soc., Washington, DC, 1976, p. 265.

210. B. A. Bierl, and E. D. DeVilbiss, "Insect Sex Attractants in Controlled Release Formulations: Measurements and Applications," in F. W. Harris, Ed., *Proc. Int. Controlled Release Pesticide Symp., Wright State Univ., Dayton, OH*, 1975, p. 230.

211. J. R. Plimmer, B. A. Bierl, R. E. Webb, and C. P. Schwalbe, "Controlled Release of Pheromone in the Gypsy Moth Program," in H. B. Scher, Ed., *Controlled Release of Pesticides (ACS Symp. Ser.* **53**), Am. Chem. Soc., Washington, DC, 1977, p. 168.

212. C. P. Schwalbe, E. C. Paszek, R. E. Webb, C. W. McComb, C. W. Dull, J. R. Plimmer, and B. A. Bierl, "Field Evaluation of Controlled Release Formulations of Disparlure for Gypsy Moth Mating Disruption," *Ann. Entomol. Soc. Am.*, 1978.

213. J. H. Tumlinson, D. D. Hardee, R. C. Gueldner, A. C. Thompson, P. A. Hedin, and J. P. Minyard, "Sex Pheromones Produced by Male Boll Weevils: Isolation, Identification and Synthesis," *Science,* **166**, 1010 (1969).

214. D. D. Hardee, G. H. McKibben, and P. M. Huddleston, "Grandlure for Boll Weevils: Controlled Release with a Laminated Plastic Dispenser," *J. Econ. Entomol.,* **68**, 477 (1975).

215. W. L. Johnson, G. H. McKibben, J. Rodriguez, and T. B. Davich, "Boll Weevil: Increased Longevity of Grandlure Using Different Formulations and Dispensers," *J. Econ. Entomol.,* **69**, 263 (1976).

216. T. B. Davich and G. H. McKibben, "Belt-wide Grandlure Formulation Test," presented at Entomol. Soc. Am. Annual Meeting, New Orleans, LA, 1975.

217. E. R. Mitchell, personal communication.

218. R. T. Staten, personal communication, 1975.

219. J. W. Peacock and R. A. Cuthbert, "Field and Laboratory Evaluations of Controlled release dispensers for *Scolytus multistriatus* pheromone, in F. W. Harris, Ed., *Proc. Int. Controlled Release Pesticide Symp., Wright State Univ. Dayton, OH*, 1975, p. 216.

220. G. T. Pearce, W. E. Gore, R. M. Silverstein, J. W. Peacock, R. A. Cuthbert, G. N. Lanier, and J. B. Simeon, "Chemical Attractants for the Smaller European Elm Bark Beetle, *Scolytus multistriatus* (Coleoptera: Scolytidae)," *J. Chem. Ecol,* **1**, 115 (1975).

221. C. R. Gentry, private communication.

222. C. R. Gentry, private communication.

223. M. Klein, USDA Japanese Beetle Research Lab, private communication.

224. F. W. Michelotti, private communication.

225. J. R. McLaughlin, E. R. Mitchell, and J. H. Tumlinson, "Evaluation of Some Formulations for Dispensing Insect Pheromones in Field and Orchard Crops," in F. W. Harris, Ed., *Proc. Int. Controlled Release Pesticide Symp., Wright State Univ., Dayton, OH,* 1975, p. 209.

226. E. R. Mitchell and J. H. Tumlinson, personal communication, 1975.

227. R. N. Sinha, "Interrelations of Physical, Chemical, and Biological Variables in the Deterioration of Stored Grains," in R. N. Sinha and W. E. Muir, Eds., *Grain Storage: Part of a System*, Avi, Westport, CT, 1973, p. 15.

228. E. J. Bond, "Chemical Control of Stored Grain Insects and Mites," in R. N. Sinha and W. E. Muir, Eds., *Grain Storage: Part of a System*, Avi, Westport, CT, 1973, p. 137.

229. B. Parker, private communication.

230. Anonymous, "Piperonyl Butoxide and Pyrethrins in the Adhesive of Cellophane-polyolefin Two-Ply Bags," *Fed. Regist.* 39(212, 38224–38225 (Oct. 30, 1974).

231. H. B. Gillenwater and L. L. McDonald, "Toxicity, Repellency and Attractancy of Slow-Release Insecticide Dispensers," *J. Georgia Entomol. Soc.,* 12(3), 261 (1977).

232. L. L. McDonald, R. H. Guy, and R. D. Spiers, "Preliminary Evaluation of New Candidate Materials as Toxicants, Repellents, and Attractants against Stored Product Insects," *USDA Mktg. Res. Rept.* 882 (1970).

233. N. L. Gauthier, "Field Experiments with Experimental Controlled Release Granular Insecticides," presented at the 1977 Controlled Release Pesticide Symp., Corvallis, OR, August 22–24, 1977.

234. A. F. Kydonieus, S. Baldwin, and S. Hyman, "Hercon Granules and Powders for Agricultural Applications," in *Proc. Int. Controlled Release Pesticide Symp. Univ. Akron, OH*, 1976, p. 4.23.

235. H. L. Collins, private communication.

236. A. Katchalsky and P. F. Curran, *Non Equilibrium Thermodynamics in Biophysics*, Harvard Univ. Press, Cambridge, MA, 1976.

237. S. Rose and J. F. Nelson, "A Continuous Long-Term Injector," *Aust. J. Exp. Biol. Med. Sci.,* 33, 415 (1955).

238. J. Urguhart, J. O. Davis, and J. T. Higgins, Jr., "Simulation of Spontaneous Secondary Hyperaldosteronism by Intravenous Infusion of Agiotensin II in Dogs with an Arterio Venous Fistula," *J. Clin. Invest.,* 43, 1355 (1964).

239. S. J. Stolzenberg, "Osmotic Fluid Reservoir for Osmotically Activated Long-term Continuous Injector Device," U.S. Patent 3,604,417 (Sept. 14, 1971); assignee: American Cynamide Company, Stamford, CT.

240. F. Theeuwes, "Elementary Osmotic Pump," *J. Pharmacol. Sci.,* 64, 1987 (1975).

241. F. Theeuwes, R. M. Gale, R. W. Baker, "Transference: A Comprehensive Parameter Governing Permeation of Solvents through Membranes," *J. Membr. Sci.,* 1, 3 (1976).

242. F. Theeuwes and S. I. Yum, "Principles of the Design and Operation of Generic Osmotic Pumps for the Delivery of Semisolid or Liquid Drug Formulations," *Ann. Biomed. Eng.,* 4, 343 (1976).

243. L. Chii-Fa, U.S. Patent 4,056,612 (1977).

244. R. H. Friedman, J. D. Krause, and W. R. Bradford, U.S. Patent 3,749,174 (1973).

245. J. G. Sayce and D. J. Brown, U.S. Patent 3,969,280 (1976).

246. A. F. Kydonieus, "Other Controlled Release Technologies and Applications," in A. F. Kydonieus, Ed., *Controlled Release Technologies: Methods, Theory and Applications*, Vol. 2, CRC Press, Boca Raton, FL, 1980, p. 242.

247. W. A. Ritschel, "Peroral Solid Dosage Forms with Prolonged Action," in E. J. Ariens, Ed., *Drug Design*, Vol. 4, Academic, New York, p. 37.

248. D. W. S. Sutherland, Ed., *Common Names of Insects and Related Organisms* (1978 revision), Entomol. Soc. Am. Spec. Pub. 78-1, College Park, MD, 1978.

CHAPTER FIVE

Microencapsulation

S. A. PATWARDHAN
National Chemical Laboratory, Pune, India

K. G. DAS
Regional Research Laboratory, Hyderabad, India

CONTENTS

1. INTRODUCTION

Microencapsulation can be described as the reproducible application of thin uniform polymeric coatings to microparticles of solids, liquids, solutions, or dispersions. The process provides an effective means to (1) convert liquids to solids for easy handling, (2) reduce toxicity of active ingredients for safety, (3) provide environmental protection to compounds that are unstable, (4) reduce volatility or inflammability of liquids, (5) mask the taste of bitter compounds, and (6) control the rate of release of active agents. These are some of the attributes derived from microencapsulation that have led to its agricultural, medicinal, pharmaceutical, veterinary, food, and feed applications. Several microencapsulated pesticides, drugs, juvenile hormones, and pheromones have been produced commercially. Microencapsulation of short-lived and highly toxic agricultural products has ensured their safe use by minimizing the serious disadvantages of high reactivity and toxicity.

Drugs such as clofibrate, vitamin E, cod liver oil, and many other liquids are converted to free-flowing pseudosolids by microencapsulation in gelatin. Aspirin microencapsulated in ethyl cellulose is a well-known dosage form.[1] The active agent acetyl salicylic acid is slowly released for 8 h. The microencapsulated form shows better gastrointestinal tolerance than other ordinary dosage forms. Microencapsulated potassium chloride shows improved gastrointestinal tolerance and better bioavailability.[2] Some of the drugs microencapsulated for palatability include vitamins and antibiotics. Microcapsules are superior to multilayer tablets in the formulation of mutually reactive materials. Aminopropylon microencapsulated with cellulose polymer can be compounded with thiamine, pyridoxine, and cobalamins.

The microcapsules range in size from several tens to a few thousand micrometers (μm). The amount of the polymeric coating can be varied from 1 to 70% by weight. Coatings ranging from 2 to 30% are recommended for commercial applications. The dry film thickness varies from 0.1 to 200 μm. Microcapsules can be isolated as dry powders or formulated in liquid vehicles for spraying. Microcapsule powders are applied with granular applicators or by suction spreaders. They are also formulated with different viscosity builders or binders in liquid vehicles for spraying. It is possible to compress or mold them into tablets, sticks, crayons, or bricks depending on the requirement and mode of application. Microcapsules ensure uniform spreading and normally have adequate shelf life.

2. CORE AND COATING MATERIALS

For microencapsulation to be successful as a controlled-release technique, there are several factors to be analyzed. The main consideration is the total system or process in which the microcapsules will be applied. Due to the lack of fore-thought there are some examples of failures in the microencapsulation approach.[3] The choice of the technique for microencapsulation that will least harm the core material to be encapsulated and the selection of the method by which the desired release mechanism can be built into the capsule wall of the coating material are very important factors for successful microencapsulation.

Water-soluble and insoluble solids, water-insoluble liquids, solutions, and dispersions of solids in liquids have been microencapsulated. Some of the agricultural and veterinary core materials that have been microencapsulated are analgesics, fumigants, antimicrobials, bacteria, disinfectants, fertilizers, fungicides, germicides, growth regulators, herbicides, insect diets, insecticides, minerals, nematocides, nutrients, repellents, pheromones, and rodenticides. Both single-particle and aggregate structures are prepared by this technique. Aggregates are composed of a number of core particles in cluster form. The particles in the aggregate need not be the same material, but all are individually coated.

The physical and chemical properties of the microcapsules depend on the nature of the coating material. The coating material should be capable of forming a film that is cohesive with the core material and should be nonreactive and chemically compatible with the core material. It should provide the requisite strength, flexibility, impermeability, optical properties, and stability to the capsule wall and be amenable to *in situ* modifications such as coloring, plasticizing, or cross-linking to build the necessary functional requirements into the microcapsule. Some of the coating properties such as solubility, cohesiveness, clarity, permeability, stability, and moisture sorption have to be considered in the correct choice of the coating material. Selection of the appropriate coating material requires knowledge of the existing literature and a study of free cast film behavior and application data.

Coating materials are selected from both natural and synthetic film-forming polymers. Some of the more common materials are aminoplasts, carboxymethyl cellulose, cellulose acetate phthalate, ethyl cellulose, gelatin, gelatin-gum arabic, nitrocellulose, polyvinyl alcohol, propyl hydroxycellulose, shellac, succinylated gelatin, and waxes.

3. PROCESSES

There are many microencapsulation techniques described in the literature. Some of the most important methods are indicated in Table 1.

3.1. Coacervation and Phase Separation

Coacervation is the phenomenon of phase separation in colloidal systems. Simple coacervation[4] results in the removal of the associated water layer from around

TABLE 1. Microencapsulation Techniques

Process	Core Material	Size (μm)
Coacervation	Solid/liquid	2–5000
Interfacial addition and condensation	Solid/liquid	2–5000
Air suspension	Solid	35–5000
Multiorifice-centrifuging	Solid/liquid	1–5000
Electrostatic deposition	Solid/liquid	1–50
Spray drying and congealing	Solid/liquid	5–600
Pan coating	Solid	600–5000

the dissolved colloid chain. This can be achieved by the addition of compounds with great affinity for water such as salts or alcohols. Such compounds compete with the colloid for the associated water molecules. When the colloid chain loses enough of its water molecules, it coacervates with other colloid chains. Figure 1 illustrates simple coacervation brought about by the addition of 20% sodium sulfate solution.

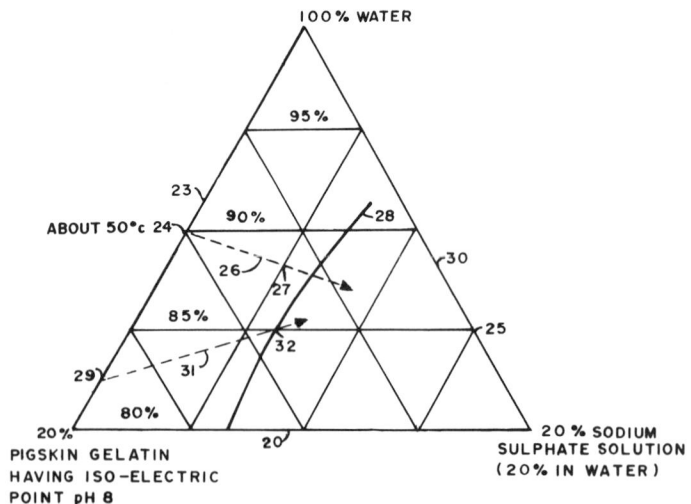

FIGURE 1. Simple coacervation.

Complex coacervation[5] (Fig. 2) involves neutralization of the charges on the

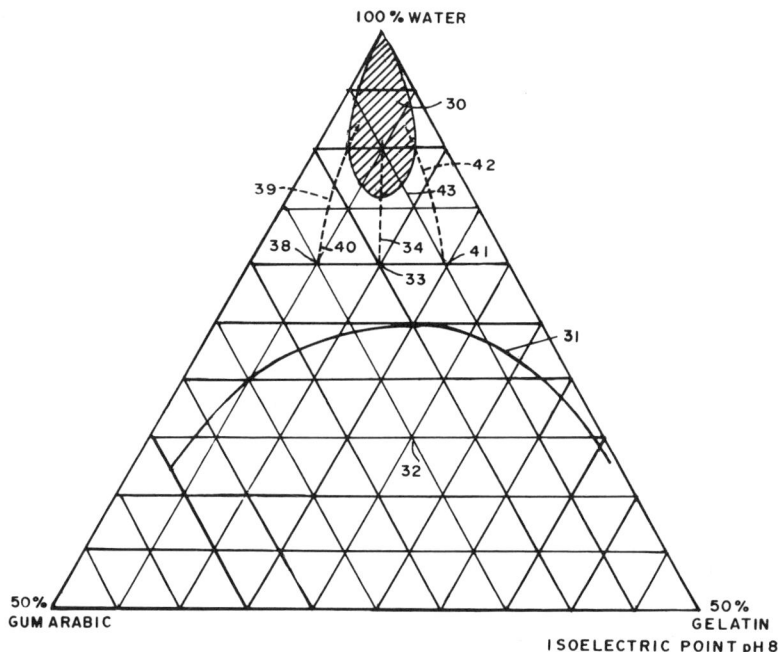

FIGURE 2. Complex coacervation.

colloid. This is effected by mixing two colloids that have opposite charges. Gelatin carries a net positive charge and gum arabic a negative charge at neutral pH; when mixed, the two colloids attract each other and separate into a distinct liquid phase called the "coacervate."

Microencapsulation by coacervation was developed by the National Cash Register Company and covered by many patents.[6] In microencapsulation the coacervate layer is deposited uniformly around the individual microparticles of the core material, which is uniformly dispersed in the medium. The nascent capsules are hardened by the addition of reagents such as formaldehyde that perform cross-linking action on the coacervate. Microencapsulation using natural[7] and synthetic[8] film-forming materials are well described in the literature.

The process consists of three steps carried out under continuous agitation (Fig. 3). The first step is the formation of three immiscible chemical phases: a vehicle, a core material, and a coating material phase. The core material is dispersed in a solution of the coating material. The vehicle is the solvent for the coating material. Phase separation or coacervation is brought about by changing the temperature of the coating material solution, by adding salt or a nonsolvent, incompatible polymer, or by inducing a polymer–polymer interaction.

In the second step, the liquid polymer coating is deposited on the core mate-

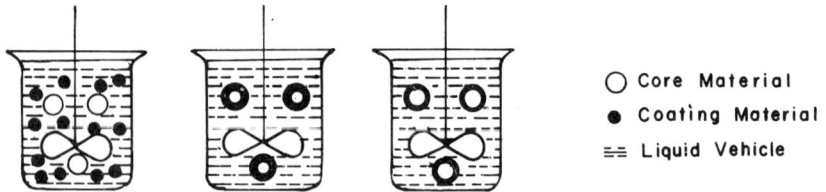

FIGURE 3.   Phase separation process.

rial. This is achieved by controlled physical mixing of the polymer in the vehicle. If the polymer is adsorbed at the interface formed between the core material and the vehicle, deposition is effective. The coating is promoted by a reduction in the total free interfacial energy of the system, resulting from the decrease in the surface area of the coating material during coalescence of the liquid polymer droplets. The final step involves solidifying the coating by thermal, cross-linking, or desolvation methods to give self-sustaining microcapsules. Subsequent spray, freeze, fluid bed, solvent, or tray drying is required to obtain the microcapsules in powder form.

3.2.   Interfacial Polymerization

Capsule wall formation has also been effected *in situ* around the core material by polymerization reactions. Brynko[9] developed a process in which the capsule wall is a solid artificial polymer material formed from a polymerized monomer material. Styrene divinyl benzene monomer was used.

Interfacial condensation reactions were successfully used for the *in situ* formation of the capsule wall. The capsule shell is formed by polycondensation of reactants situated respectively at opposite sides of the interface between droplets that initially constitute the material to be encapsulated and a surrounding continuous phase of liquid. Typical polycondensate capsule shells are polyamide, polysulfonamide, polyester, polycarbonate, polyurethane, and polyurea.

The reaction between an aqueous solution of a diamine and an organic phase containing the diacid chloride has been utilized for microencapsulation by interfacial polymerization. The active ingredient to be encapsulated is also dissolved in the organic phase. Stabilized emulsions of the oil droplets containing the diacid chloride in the water phase are prepared using low concentrations of nonionic surfactants and high-shear stirrers. Addition of the diamine to the aqueous phase of the emulsion leads to instantaneous polymerization at the interface of each oil droplet to form a film completely enclosing the droplet containing the active ingredient. Polymerization proceeds, since the diamine diffuses through the polymer to react with all the diacid chloride. An acid acceptor present in the aqueous phase helps remove the HCl formed and makes all the diamine available for polymer formation (Fig. 4).

The microcapsules formed are filtered off and dried to give free-flowing powders containing as high as 90% liquid payload of the active agent. To prepare sprayable formulations, the sieving step and a stabilization treatment with a thickening or suspending agent are necessary.

Organic liquids (low-melting solids) with low water solubility and without any functional groups that are likely to react with the encapsulation systems are amenable to this technique. Insoluble solids are handled as slurry in an organic liquid phase. Water-soluble materials offer many practical problems.

Cross-linking of the polymer wall is desirable to obtain durable and stable capsules.[10] Polyfunctional reactants are included in polycondensation mixtures. The wall thickness is a function of the ratio of the quantity of polymer to the encapsulate and the capsule size. The efficacy, toxicity, and cost of production depend on capsule wall thickness. The particle size is equally important. The relationship between capsule size and wall thickness, assuming the density of the polymer wall and encapsulate is equal to 1.0, can be expressed as:

$$W = r_1 - r_2$$

where w = wall thickness, r_1 = capsule radius, r_2 = radius of encapsulate.

$$W = \sqrt[3]{\frac{3V_1}{4\pi}} - \sqrt[3]{\frac{100 - \% \text{ wall}}{100} \left(\frac{3V_1}{4\pi}\right)}$$

where V_1 = volume of capsule.

The morphology of a capsule depends on what it contains. The encapsulate can change the characteristics of the interface between the two liquid phases participating in the polymerization. Capsule morphology also determines release properties and interactions of the capsule and substrate. Rough capsule walls have more surface area, thin spots, and imperfections than smooth ones and would contribute to faster release. Spraying microencapsulated formulations at

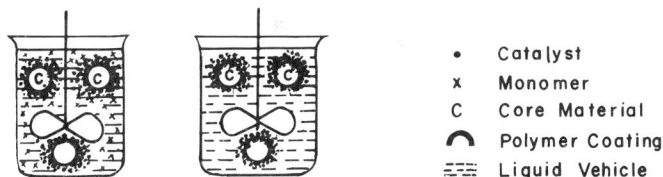

FIGURE 4. Interfacial polymerization.

- •   Catalyst
- x   Monomer
- C   Core Material
- ⌢   Polymer Coating
- ≡≡≡   Liquid Vehicle

the same active ingredient rate as aqueous emulsions does not cover 100% of the area. For example, 27-μm microcapsules give 0.6% coverage. However, unlike standard aqueous emulsions, the encapsulated active agent is localized at higher concentrations and hence is not prone to adsorption by the substrate. Ultraviolet degradation and chemical reactions are avoided by the protection offered by the capsule wall. For 27-μm capsules, at a spray rate of 1 lb of active agent per acre, the center-to-center capsule spacing will be 0.3 mm. Thus with less than 1% area coverage the efficacy will be appreciable. Capsules 10-100 μm in size are most useful because of their high efficiency, low tendency to drift, increased residual activity, and less inhalation hazard.

Penncap-M insecticide manufactured by Pennwalt Corporation is encapsulated methyl parathion which has been commercialized.[11,12] A few developmental and experimental products[13,14] are under evaluation (Fig. 5).

3.3. The Wurster Process

In a process invented by Prof. D. E. Wurster,[15] the particles to be coated are fluidized on an upward-moving airstream (Fig. 6). A cyclic flow of the particles is produced. They are accelerated and physically separated when they enter a high-velocity spout. The coating material is applied through a spray nozzle mounted at the base of the spout. Since smaller particles have more surface area per unit weight, they require more coating to achieve the same level of en-

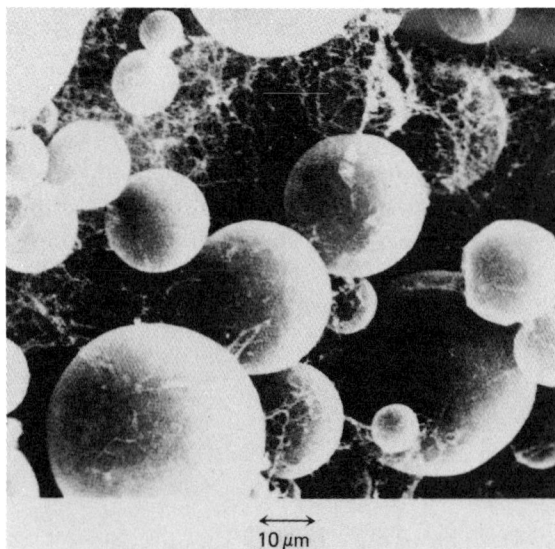

← 10 μm →

FIGURE 5. Microencapsulated synthetic pyrethrum.

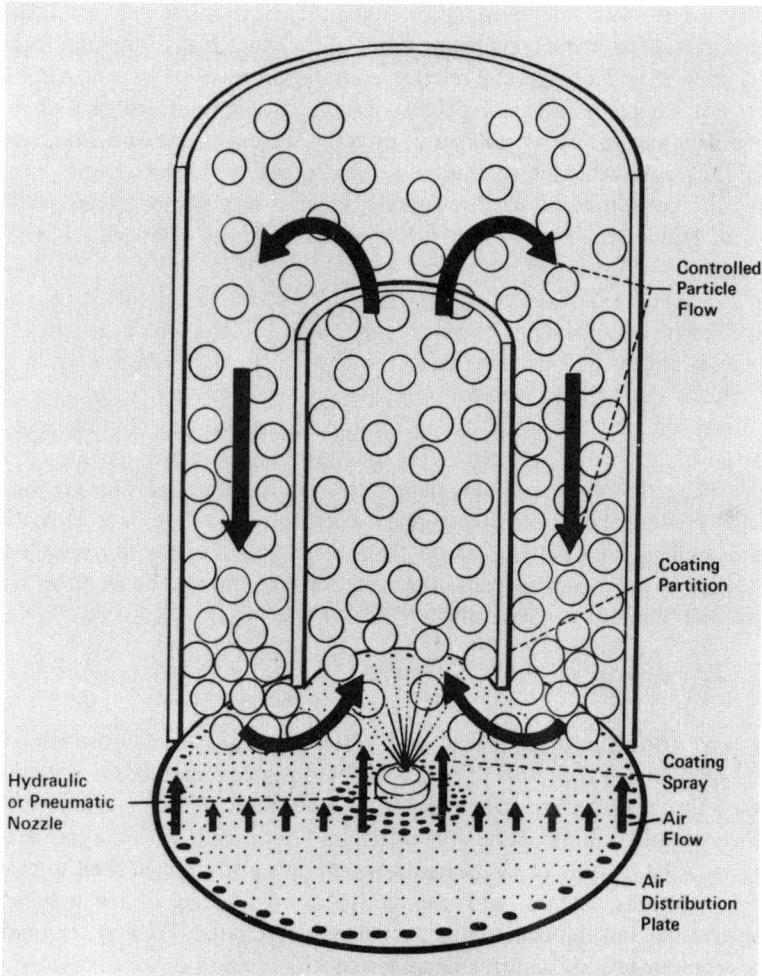

FIGURE 6. Wurster coating chamber.

capsulation. A limiting factor in encapsulating smaller particles is the size of the atomized droplets of the coating solution. If droplets are large, the particles tend to stick together, and the removal of the solvent becomes slow which also enhances agglomeration of particles. This method is used when the particle size of the coated product is larger than 140 mesh (160 μm).

The properly selected solvent is a very important component of any coating system.[16] Fast-drying solvents are preferred to obtain good drying conditions. Solvents and solvent blends are selected with the help of the available data on solubility parameters and evaporation rates. Hydrogen bonding and dipole moments also affect the drying properties.[17] The interaction of the film former and the solvent will change the relative evaporation rates of solvents. Boiling point is not a reliable indicator of evaporation rate. Advantages such as cost, safety in handling, safety of residues in product, solvent power of the resin, and compatibility with other ingredients have to be considered in the selection of the solvent. The core material and the coating material may be heat sensitive. The process air temperatures may be in excess of the supposed temperature limit in some cases.

The process air that fluidizes the particles also serves to dry the coating, and the particles are completely dried by the time they clear the top of the partition. When the airstream and particles clear the top of the partition, the air in the spout spreads out to fill the expansion chamber. Once the air spreads out and slows down, the particles settle out to the top of the bed and then descend to the botton of the bed. They reenter the partition, are fluidized, and are coated again. Particles thus cycle and pass the nozzle every 6-10 s, each time getting an additional coating. The process is stopped when the desired amount of coating has been applied. The particles range from a few micrometers to large pieces. They can be cubes, noodles, ovals, disks, or irregular forms. The air flow, temperature, and humidity are controlled.

3.4. Physical Methods

Nozzle-type devices have been used on a laboratory scale for encapsulation by physical methods. The common feature of these physical methods is the use of fluid shell formulations and hardening of the shell after capsule formation. The fluid shell formulation should be nonreactive and immiscible with the core material and must be capable of easily hardening to give a film. Such shell formulations are solutions, latexes, and hot melts. A cross-linking of the polymeric coating material and the core component may be tolerated. To meet the immiscibility requirement, an aqueous coating material solution or a latex would be appropriate to encapsulate a water-insoluble material. A nonaqueous shell solution can be used to encapsulate aqueous solutions. If the core and coating materials are handled as hot molts for encapsulation, the immiscibility requirement may not be very important, since before any considerable amount of intermixing occurs the shell will be hardened.

The nozzle-type encapsulation processes and equipment were initially developed at Southwest Research Institute.[18] The fluid shell material is admitted to a gravity-flow nozzle. It flows through the annular space to a circular orifice located at the bottom of the device and forms a membrane. Through a central tube the encapsulating material is fed dropwise onto the membrane. The compound fluid droplet formed falls from the nozzle to an appropriate hardening bath. Surface tension causes the fluid capsule to assume a spherical shape before it falls into the bath. A new membrane is formed across the orifice as soon as the previous capsule separates from the nozzle, and the next capsule is then formed. These processes are continuously repeated.

Many commercially available synthetic polymers and natural waxes, resins, and gums meet the general requirements of shell materials. Mixtures of shell materials are also used. Gelling agents are added to provide initial structural strength and to get the desired shell properties. Cross-linking on the shell polymer has been found to control the swelling of moisture-sensitive capsule shells. Release characteristics can change with the nature and stability of the shell material. Plasticization can be advantageous in some cases.

The core material is pumped through nozzle devices. Liquids are best suited for encapsulation, but finely divided solids slurried in a liquid have also been encapsulated. Such slurries can help in isolating the solid particles from the fluid shell until the shell is hardened. Thermally stable core materials can be encapsulated in the molten state as liquids.

The choice of shell-hardening technique is based on the shell formation method employed. The common processes are based on chemical reaction, solvent extraction or evaporation, cooling, and their appropriate combinations. Freshly formed capsules of sodium alginate or sodium polypectate are hardened with a calcium salt solution. The capsules are rinsed, and the residual water is removed by fluidized bed drying. Hardening by solvent removal is effected by evaporation or by receiving the freshly formed capsules in a suitable nonsolvent. The solvent–nonsolvent system should not precipitate the polymer, and the shells should remain separated during solvent extraction. Cooling is another method of shell hardening.

Shell solutions and latexes afford excellent means of preparing high capsule payloads. The shells shrink during solvent removal. The optimum shell formulation viscosity for the nozzle device process can be obtained by either selecting a polymer within a suitable molecular weight range or by adjusting the total solids. A compatible gelling agent is often added to get the required total solid content.

Many developments have given new dimensions to this approach. The use of a multiorifice centrifugal head is one of the early improvements. The main part is a head that revolves on a vertical axis. Two streams of the fluid shell material enter the head, and each stream flows into an internal groove. Centrifugal forces hold the material until it overflows an internal weir and flows into an area containing a row of counterbored orifices. Membranes are formed across each orifice. The core material is atomized and transferred to the membrane by means of a concentrically rotating disc. Capsule formation takes place when the combined mass of core and shell material becomes sufficient for centrifugal force to overcome the cohesive force of the shell material. The capsule is projected outwards from the periphery of the head. The capsule size is directly proportional to the rotational speed of the head.

Capabilities for controlling capsule size and uniformity and for increasing the rate of production were built into many of the devices developed later. The centrifugal encapsulation technique was combined with fluid extrusion. Figures 7 and 8 show simple extrusion and multiorifice centrifugal extrusion devices. A hardening bath with verticle baffle is illustrated in Fig. 9.

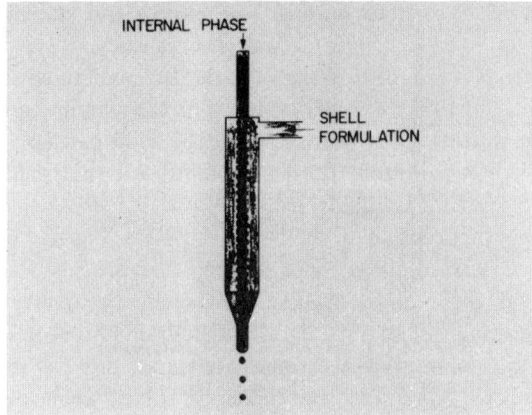

FIGURE 7.   Simple extrusion.

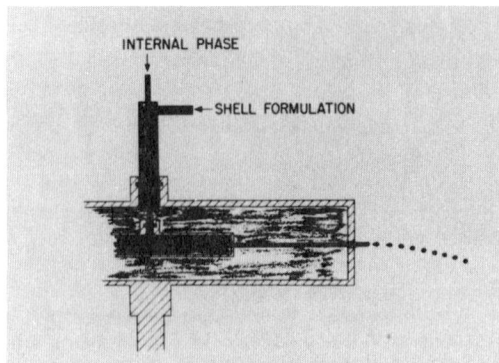

FIGURE 8.   Multiorifice centrifugal extrusion head.

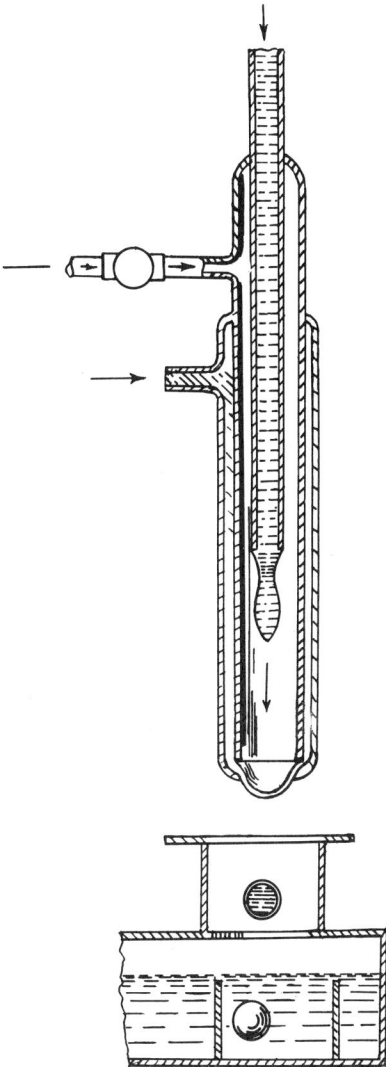

FIGURE 9. Hardening bath with verticle baffle.

The core material is pumped through the central tube, and the shell material through the annular space. A fluid "rod" of the core encased in a sheath of shell material is thus extruded. At some point beyond the tip of the nozzle, the fluid rod breaks into individual fluid capsules. Since the uniformity of the capsules is poor and size control is limited, centrifugal extrusion heads, submerged extrusion devices, and vibrating centrifugal extrusion nozzles were developed. Capsules

350 μm in diameter have been prepared at rates exceeding 300,000/s per orifice. The scale-up potential of these devices has been demonstrated.

4. PRODUCTS

4.1. Experimental

Mirex, a fire ant toxicant encapsulated as a soybean oil solution has been reported[19] to be as effective as a standard commercial form. A dose of 100 g/acre was recommended, compared to 565 g/acre for the standard commercial formulation. Each form supplied the same amount of the active ingredient per acre. The microencapsulated bait showed increased field persistence and was found not only to eradicate the old mounds but also to control the formation of new mounds. This is an example of effectively converting a liquid to solid for efficient use.

Lepidoptera pheromones are highly expensive and volatile substances. To disrupt lepidoptera mating activities, only very small quantities of the pheromones are needed for vast acreage. Their population buildup can be controlled if a minimum vapor concentration of the pheromone can be maintained in the air over the host crop for the entire mating season. The economic damage to the crop by the insect can be thus minimized. Beroza and co-workers[20] have demonstrated the potential of microencapsulated disparlure for control of gypsy moth, *Porthetria dispar* (L.). The product was effective for 6 weeks. The results on large-scale field trials were not encouraging. Microencapsulated pheromone of the redbanded leaf roller, *Argyrotaenia velutinana*, (Wlk.), a major apple pest, showed 75–99% disruption of the attraction of males to monitoring traps during field trials. Mixtures of microencapsulated trimedlure, the pheromone of Mediterranean fruit fly, *Ceratitis capatata* (Wiedemann), with microencapsulated insecticides are under investigation.[21] Microencapsulated (Z)-7-dodecen-1-ol acetate, the cabbage looper pheromone, was tested by spraying on cabbage. The capture of male cabbage loopers in traps were reduced by 80% in 24 days, indicating that mating would have been reduced by the same amount during that period.[22]

Microencapsulated gardona for the control of fly larvae in cow manure and encapsulated rabon for house fly larvae control in manure were reported to be effective.[23] Microencapsulated chlorpropham was reported to show herbicidal properties for a longer period than the commercial formulation.[24] The loss of active agent by evaporation was also considerably reduced. Urea, acetylsalicylic acid, phenylbutazone, and microbial materials have been microencapsulated to increase flow properties, reduce toxicity, and mask taste.

The Wurster process has been applied to coat pharmaceutical tablets, to encapsulate pediatric drugs, to prepare enteric coated drugs, and to make oral vaccines. In the food industry, vitamins and minerals have been encapsulated.

The Wurster process has been used to coat cotton seeds with a polymer containing disyston and onion seeds with fungicide to enhance the efficiency of the pesticide by its controlled release. Seeds have also been coated to increase their size to permit machine planting. Catalysts, some salts, and metallic particles have been coated to incorporate specific desired properties, such as delayed action, for separation of incompatibles and to give a desired form to the particles. A controlled-release mosquito larvicide has been prepared.[25]

Among drugs, cod liver oil, castor oil, clofibrate, vitamin E, and triglycerides have been microencapsulated in hard gelatin capsules by coacervation techniques. Potassium chloride, flucloxacillin, antibiotics, and aminopropylon have also been microencapsulated. As a result of microencapsulation, liquid forms have been converted to pseudo-solids, the gastrointestinal irritation caused by some drugs taken orally can be avoided, and objectionable tastes and odors have been masked.

Dan Fraenkel of the Weizmann Institute of Science, Israel, has encapsulated hydrogen in zeolites.[26] $C_{S2.6}$–A zeolite has been reported to be the best storing system for encapsulating hydrogen gas. At 200°C and 600 atm for 2 h, a $C_{S2.6}$–A sample held under ambient conditions up to 0.8 wt % of hydrogen. Zeolite/hydrogen encapsulates have great potential as energy storage media. The main disadvantage is the dependence on costly and scarce high-pressure technologies. Zeolite encapsulates of gases and reagents have great potential in maintaining optimum reactant concentrations and improving catalytic behavior.

4.2. Commercial

Penncap-M insecticide, a microencapsulated methyl parathion, is the first encapsulated product for agricultural application distributed commercially by the Pennwalt Corporation.[12] The product was granted registration for use on 16 major crops. A cross-linked polyamide-polyurea polymer provides the capsule wall. The wall formation takes place when sebacoyl chloride and polymethylene polyphenylisocyanate react with a mixture of ethylenediamine and diethylenetriamine at the interface of microdroplets of methyl parathion dispersed in water.

Equi-palazone is a taste-masked phenylbutazone marketed by Arnold's Veterinary Products Ltd., Reading, England. It is a taste-masked microencapsulated antiinflammatory agent marketed for horses in various parts of the world.

Triglicer is microencapsulated clofibrate manufactured by Medicamehta Company, Lisbon, recommended for the treatment of hypercholesterolemia and hyperlipemia, conditions of excess cholesterol and fat, respectively, in the blood. It is an oily liquid with an unpleasant taste and odor. Microencapsulation converts it to a dry free-flowing powder consisting of microscopic granules of uniform size, individually coated with an ultrathin membrane made of pharmacologically inert and completely harmless material. This pseudo-powder can be filled into hard gelatin capsules. The capsule is 75-90% active core material and 10-25% gelatin, the membrane material. About 95% of the microcapsules are less than 500 μm in diameter.

Encapsulated Warfarin prepared by the Wurster process is the first rodenticide registered with the Environmental Protection Agency. Warfarin is encapsulated to mask taste and odor, and the active agent is released in the gut of the animal. Since the coating prevents the rodent from detecting the presence of the toxin, it will eat more of the toxic bait and thus the efficacy of the toxin is greatly improved.

Precise Timed Release plant food and African violet food are two products marketed by the 3-M Company. With one application, plant food lasts for 3-4 months and African violet food for 4-6 months. These products are used as fertilizers for home use.

4.2. Developmental

Microencapsulated ethyl parathion Penncap-E insecticide is being evaluated under an experimental permit for use against some insects on sorghum. Microencapsulated diazinon identified as Knox Out 2 FM insecticide is also being evaluated under an EPA experimental permit.

Tables 2-4 present data on the number of microcapsules per gram per square meter and the stability of encapsulated solvents.

5. PARAMETERS AFFECTING PROPERTIES

The coating material is a major determinant of the physical and chemical properties of the microcapsules. The coating polymer should be capable of forming a thin film, chemically nonreactive, capable of providing the desired coating prop-

TABLE 2. Number of Microcapsules per Gram
of Material with Various Diameters

Diameter (μm)	Microcapsules per Gram
5	15,279,000,000
50	15,279,000
100	1,909,800
500	15,277
1000	1,910

TABLE 3. Microcapsule Quantities according to Capsule Size and Distribution Rate

Mass Distribution (g/ha)	Capsules per Square Meter			
	5 μm	50 μm	100 μm	500 μm
1	1,527,900	1,530	191	1.5
50	76,395,000	76,395	9,550	75
100	152,790,000	152,790	19,098	152
200	305,580,000	305,580	38,196	304
400	611,160,000	611,160	76,392	608

TABLE 4. Stability of Encapsulated Solvents

	Solvent in Capsule (%)	Capsule Size (μm)	Days	Solvent Loss at 77°F, 50% RH
Benzene	85.5	500	198	0.5
Carbon tetrachloride	82.8	500	602	0.3
Chloroform	78.9	420	730	0.1
Hexane	70.8	35	730	0.1
Toluene	89.4	20	600	0.1
Xylene	90.2	500	730	0.2

erties, and amenable to *in situ* modifications. The coating may be plasticized or chemically altered by cross-linking to give the desired dissolution or permeability.

For interfacial polymerization techniques, the choice should be complementary organic polycondensate-forming intermediates such as a diamine or diacid chloride. Polyamides, polyureas, polyesters, polysulfonamides, and polycarbonates can be produced by interfacial polymerization. To prepare capsules with satisfactory shelf life and durability, it is desirable to cross-link the polymer wall. A polyfunctional reactant is generally added to the medium for this purpose. The microcapsule wall thickness is a function of the ratio of the amount of polymer formed to the quantity of encapsulate. It is also a function of the size of the capsule. Thick-walled capsules are uneconomical to prepare and carry less payload even though they are more durable and secure. Thin-walled capsules are less costly to produce and carry greater payloads, but offer less protection to the active agent and are fragile. The capsule wall thickness affects the cost of production, efficacy, and toxicity.

Another important variable is particle size. Large particles are difficult to spray, and very small capsules can break easily. A 27-μm capsule with a wall 10% by weight will have a wall thickness of 0.5 μm and a capsule volume of 10^{-8} ml. One gram of the product will contain 10^8 capsules, and the capsule payload will be 90%.

6. RELEASE MECHANISMS

In membrane-encapsulated reservoir devices, a rate-controlling membrane encloses the active agent and determines the release rate. The membrane may be porous or nonporous. When the reservoir contains a suspension of the active ingredient in a fluid, it is easy to maintain a constant activity of the agent in the depot until the excess is removed. The ideal situation of a constant steady-state release rate by diffusion through the membrane is possible.

If s is the surface area, D the diffusion coefficient of active agent in the membrane, C_s^m the solubility of active agent in the membrane, and l the membrane thickness, the release rate can be expressed as $SD\,C_s^m/l$. If a solution of the active ingredient in the reservoir fluid is encapsulated, the activity will change and the release rate will decrease more or less exponentially with time and follow first-order kinetics. The duration and rate can be engineered by the design of the system. Encapsulated systems also release the active agent by mechanical rupture of the membrane.

Many factors affect membrane permeability. Even though the transport properties of homogeneous polymers are well established and predictable, the transport behavior of heterogeneous polymer membranes is more complex and less predictable. These membranes show great promise for unique permeation and release processes. The pure diffusive flux in the steady state can be expressed as

$$J = -DK\frac{dc}{dx}$$

The diffusion coefficient D^* is a measure of the average mobility of penetrant molecules within the diffusion medium. The apparent Fick's law diffusion coefficient is the product of the non-negative mobility factor and a thermodynamic factor related to the ideality of the penetrant–polymer mixture.

$$D = D^*\frac{d\ln a}{d\ln \bar{c}}$$

where a is the activity and $\bar{c}$ is the concentration of penetrant in solution in the polymer phase. When the mixture is a thermodynamically ideal solution, $(d \ln a)/(d \ln \bar{c})$ is unity. Most mixtures exhibit deviations from ideal behavior of a magnitude proportional to the concentration. The concentration dependence of D arises from the concentration dependence of the mobility and the nonideal nature of the system.

If for any reason the thermodynamic terms become negative in sign, diffusion may occur against the concentration gradient. The presence of unstable phase regions within the diffusion medium due to sudden local changes in temperature, composition, morphology, or applied stress may lead to "uphill" diffusion.

The distribution factor is a measure of the partitioning of active agent between a polymer solution phase and the external phase to the membrane. This Nernst-type distribution function of a penetrant–polymer system can be a function of pressure, concentration, or temperature. The product of the mobility and distribution parameters may be defined as the permeability. Under the influence of a concentration gradient, the molecules of the active agent move in unit steps by a cooperative action of the surrounding complex from one position to the next. The physical and chemical properties of the components and the experimental conditions govern the relative mobilities of the active agent molecules and polymeric chain segments. Hole formation depends on the size, shape, concentration, and interaction between components. The number, size, and distribution of voids, capillaries, and domain boundaries within the polymer membrane also affect transport.

Chain interactions due to hydrogen bonding, polar groups, or van der Waals forces in the polymeric membrane reduce mobility and permeation rates. A high degree of cross-linking, the presence of solid additives, and the occurrence of crystalline regions within the polymer decrease chain segmental mobility and release rates. The incorporation of plasticizer and permeation of solvents in a polymer enhance local chain segmental mobility.

For low concentrations of the active agent, where Henry's law is valid, the diffusion coefficient varies with concentration. With wider sorbed concentrations where Henry's law is not obeyed, the sorption follows the Flory-Huggins equation. Theories have been proposed to explain the observed concentration dependence of diffusion in terms of free volume concepts and to rationalize the phenomenon of the formation of active agent clusters. The assumption has been made that the accessible regions are structurally homogeneous and diffusion occurs by a single activated mechanism. The presence of microporous structure in certain amorphous polymers below or near their glass transition temperature has been well established.

The distribution of void, size, and shape depend on the mode of membrane preparation. The polymers may have unit-cell dimensions or porosities and cleavages of greater dimensions and nonrandom configuration. They have to be distinguished from the free volume associated with liquids or amorphous solids. Their magnitude is not a thermodynamic quantity. Microporous structures affect solubility and transport properties. The presence of interconnected micropores permits convection of active ingredients through the polymer in addition to activated diffusion.

Modification of polymer structure and properties can alter release characteristics. Polymer structures are modified by changing composition, morphology, and geometry. Graft and block copolymer systems allow independent changes in polymer morphology and composition.

REFERENCES

1. V. H. Saggers, L. F. Chasseaud, and A. J. Cooper, *Clin. Trials J.*, 9,36(1972).

2. G. C. Maggi and G. Coppi, *Curr. Ther. Res. Clin. Exp.*, 21, 676 (1977).

3. J. E. Vandegar, *Proc. 5th Int. Symp. Controlled Release Bioactive Materials, Gaithersburg, MD*, 1978, p. 2.1.

4. B. K. Green and L. Schleicher, U.S. Patent, 2,800,457 (July 23, 1957).

5. B. K. Green, U.S. Patent Reissue 24,899 (1960).

6. J. A. Bakan "Microencapsulation using Coacervation/Phase Separation Techniques," in A. F. Kydonieus, Ed., *Controlled Release Technologies: Methods, Theory and Applications*, CRC Press, FL, 1980, p. 84.

7. J. A. Bakan, U.S. Patent 3,567,650 (Mar. 2, 1971).

8. E. H. Jensen and J. G. Wagner, U.S. Patent 3,069,370 (Dec. 18, 1962).

9. C. Brynko, U.S. Patent 2,969,330 (Jan. 24, 1961).

10. J. R. Lowell, Jr., W. H. Culver, and C. B. De Savigry, in H. B. Scher, Ed., *Controlled Release Pesticides (ACS Symp. Ser. 53)*, Am. Chem. Soc., Washington, DC, 1977.

11. E. E. Ivy, "Penncap-M," *J. Econ. Entomol.*, 54, 473 (1972).

12. Anon., *Chem. Eng. News*, July 29, 15 (1974).

13. M. Beroza, L. J. Stevens, B. A. Bieri, F. M. Phillips, and J. G. R. Tardif, *Environ. Entomol.*, 2, 1051 (1973).

14. R. C. Koestler, "Microencapsulation by Interfacial Polymerization Techniques—Agricultural Applications," in A. F. Kydonieus, Ed., *Controlled Release Technologies: Methods, Theory and Applications*, CRC Press, Boca Raton, FL, 1980, p. 127.

15. D. E. Wurster, U.S. Patent 3,253,944 (1966).

16. H. Hall and T. Hinkes, "Air Suspension Encapsulation of Moisture Sensitive Particles Using Aqueous Systems," in J. E. Vandergaer, Ed., *Microencapsulation: Processes and Applications*, Plenum, New York, 1974, p. 145.

17. J. D. Crowley, G. S. Teague, Jr., and J. W. Lowe, Jr., *J. Paint Technol.*, 38, 269 (1966).

18. C. F. Raley, W. J. Burkett, and J. S. Swearingen, U.S. Patent 2,766,478 (1956).

19. G. Markin and S. Hill, *J. Econ. Entomol.*, 61, 193 (1971).

20. M. Beroza, C. S. Hood, D. Trefrey, D. E. Leonard, E. F. Knipling, W. Klassen, and L. J. Stevens, *J. Econ. Entomol.*, 67, 659 (1974).

21. I. Keiser, R. M. Kobayashi, and E. J. Harris, in F. W. Harris, Ed., *Proc. Int. Controlled Release Pesticide Symp., Wright State Univ., Dayton, OH*, 1975, p. 264.

22. J. R. McLaughlin, E. R. Mitchell, and J. H. Tumlinson, F. W. Harris, Ed., *Proc. Int. Controlled Release Pesticide Symp., Wright State Univ., Dayton, OH*, 1975, p. 209.

23. R. W. Miller and C. H. Gordon, *J. Econ. Entomol.*, 65, 455 (1972).

24. B. C. Turner, D. E. Glotfelty, A. W. Taylor, and D. R. Watson, *Abstr. Papers, 173rd Am. Chem. Soc. Nat. Meet., New Orleans, March 1977*, Paper 8, Pesticide Div.

25. K. G. Das and V. B. Tungikar, in R. W. Baker, Ed., *Proc. 6th Int. Symp. Controlled Release Bioactive Materials, New Orleans*, 1979, p. IV-37.

26. D. Fraenkel, *Chem. Technol.*, 1981, 60.

CHAPTER SIX

Determination of Release Rates

R. M. WILKINS
Department of Agricultural Biology
The University of Newcastle upon Tyne
Newcastle upon Tyne, United Kingdom

CONTENTS

1. GENERAL PRINCIPLES

Controlled-release formulations for use in agricultural and forestry applications can be distinguished from conventional formulations mainly in the rate at which the active ingredient is made available to the target organism. Determination of this release rate is thus the crux of the development of efficient formulations and a more rational delivery of agrochemicals.

Determination of release rates can have two important functions. The first is the characterization of a particular formulation to ensure that the release-rate profile is appropriate to the requirements of the pest control response. In most cases a knowledge of the required dosage–response relationship in the field is lacking. In some situations this lack of information may be overcome by simulation, under field conditions, of various application regimes, such as frequent repeated sprays.[1] This approach can help where the nature of the response is initially undefined, for example, in the development of controlled-release plant growth regulator formulations.[2] Controlled release of herbicides has provided growth stimulation to seedling conifers[3] and has also given aquatic weed control at extremely low rates of release, the "chronicity effect,"[4] both of which were unforeseen. Thus, in practice, assumptions are usually made regarding the required rate of delivery, and the formulation is then evaluated on the basis of subsequent performance.

The second function of the determination of release rates is to provide a theoretical basis for the rational design of formulations based on a knowledge of the rate-controlling mechanism.[5] Such studies are carried out under rigorously controlled conditions. It is impossible to reproduce environmental conditions in the laboratory, and in any particular environment the rate-controlling step may be different from that determined under laboratory conditions. Boundary diffusion may be important according to the environment,[6] as may partition of the active ingredient between the formulation and the surroundings.[7] These factors will vary considerably in the natural environment, and an excellent account of their effect on chemical movement is provided by Hartley and Graham-Bryce.[8] In addition, the variability of the crop environment can cause different mechanisms to be operative in the same formulation, particularly in the soil. The importance of this variability, especially in chemically bonded systems, has been emphasized by Allan et al.[3,9]

Experimental determination of release rates of formulations for agricultural and forestry application can be conveniently divided into three categories. At the simplest level are the *in vitro* tests carried out under abiotic, controlled environmental conditions. These usually involve release into water or air and allow the experimenter to control the concentration gradient across the formulation-medium interface and to obtain detailed release rates with convenient analytical techniques. At the other end of the scale are release rate determinations made under field conditions. These are more expensive in terms of effort and materials and need considerable replication to enable reliable predictions to be made.

An intermediate category of testing between these two extremes can be devised, combining elements of both. This approach includes aquatic testing in small vessels containing soils, plants, and animals and soil-containing pots in controlled environmental or greenhouse conditions. Some control under biological conditions is thus possible, and consequently release rates determined in this way are more indicative of field performance than *in vitro* assays.

The relevance of release rates determined by any particular experimental method depends on the type of formulation, whether chemical or physical release mechanisms are involved, and also on the medium for release. For agricultural applications, the intended medium, which provides the means for active ingredient delivery to the target organism, is either water, air, soil, or actual contact with the surface of the target itself. The soil environment, the most important of these for agricultural chemicals, contains elements of the others and thus presents problems for characterization of formulations.

The aim of this chapter is to review the methodology employed for the characterization of controlled-release formulations for agricultural applications. The experimental approach has been based on whether the formulation is intended for delivery to (1) air or vapor phase, (2) aquatic situations, or (3) soils or plants. This classification obviously leaves out some areas, including veterinary applications and release to inert surfaces to maintain a bioactive depot. Where evidence permits, a comparison will be drawn between release rate determinations made under varying experimental conditions as outlined above. For a discussion of controlled-release formulations, their preparation, and mechanisms of release, the reader is directed to other chapters of this volume.

1.1 Characterization and Analysis of Controlled-Release Formulations

An essential part of the characterization of any controlled-release formulation is the total releasable content of the active ingredient. It is thus desirable that the amount of the active agent be measured before use by an appropriate technique independent of that estimated from the proportion of ingredients employed in the preparation. A distinction must be made between the active ingredient content and the releasable or available active ingredient. A further complication is created by the fact that formulations in which the release rate is slow enough to appear pseudo-zero-order for appreciable periods[10] release only a portion of the available active ingredient before the rate drops to levels insufficient to maintain the required bioactivity.

Thus to obtain a true analysis of the releasable active ingredient content, a reversal of the preparation of the formulation is required. This is in effect an acceleration of the normal release for which the rate constant is inversely dependent on the free energy of formation of the bond between the active agent and the substrate. To understand this, we could consider all the formulations in terms of thermodynamics.[11] Thus all controlled-release formulations can be regarded as part of a spectrum of bonded pesticides ranging from the low energy of adsorption on appropriate substrates to the high bonding energy in covalent bonds. Formulations in which the pesticide is dispersed or encapsulated in polymers can also be included in this spectrum.[12] In these cases, the pesticide, at some stage, interacts with the polymer to form a solid solution. The low energy of the individual interactions that delay release are supplemented by the additional constraints of diffusion through the matrix. These thermodynamic considerations logically apply to all formulations that release by chemical or physical mechanisms.

Except where analysis of the formulation can be carried out with the formulation itself, usually by some form of spectrometry or neutron-activation analsis,[10] the active ingredient must be separated from the inert substrate. Where low-energy bonding is involved, simple solvent extraction is sufficient. For rubber septa impregnated with a volatile pheromone, shaking with hexane-dichloromethane (1:1) for 1 h was employed.[13,14] For laminated plastic dispensers an overnight soak in hexane[15] was required, and for microencapsulated pheromones more energy was needed in the form of grinding in a mixer-mill with a ball bearing prior to extraction with acetone.[16] Routine analysis of commercial microencapsulated diazinon, ethyl parathion, and methyl parathion has been based on grinding and extraction into acetonitrile[17] or extraction with hexane or ethyl acetate.[18] Extraction of volatile pheromones from plasticized PVC was achieved by immersion in ethyl acetate for 10 h,[19] while aldicarb needed refluxing in chloroform for 4 h for separation from polyurethane, PVC, cellulose acetate, cellulose acetate butyrate, polyester, or urea-formaldehyde.[20] The aldicarb content of polyamide granules was determined by completely dissolving the granules in methanol and analyzing the mixed solution using colorimetry.[20] Separation can also be based on the low volatility of polymers compared to the active ingredient. This can be exploited by direct injection of a solution of the mixture into the precolumn of a gas chromatograph or into a heated tube containing glass wool in a sweep codistillation apparatus.[10]

Analysis of chemical combinations requires reaction conditions appropriate to the bond energy of the pendent pesticide. With the most common[3,21] ester or amide-linked carboxylic acid pesticides, a convenient method is saponification with alcoholic potassium hydroxide.[22] Thus, following removal of any free acid from the formulation using solvent extraction,[23] the content of releasable active ingredient can be determined.[24,25] This can apply to insoluble polymer substrates[23] as well as to water-soluble backbones.[26,27] A range of pesticides has been linked to macromolecules,[21] and some have been characterized by bioassay, but little attention has been paid to quantifying their releasable active ingredient content. For descriptions of analytical methods for the pesticides, the reader is directed to general references[28,29] or to the literature for methods specific for the chemical under consideration.

Other aspects of the characterization of formulations can be demonstrated using scanning electron microscopy. Thus the structure of microcapsules can be related to the rate and mechanism of release.[30] This method can be extended to matrix formulations, and relative rates of release of thiocarbamate (EPTC and butrate) herbicides from starch xanthide granules have been correlated with surface and internal structure as shown under the scanning electron microscope.[31]

This method can be further used to help to determine mechanisms of release. Kraft lignin provides a biodegradable polymer[32] that is compatible with certain pesticides.[33,34] Formulations may be prepared in which the lignin is plasticized by a pesticide and formed into granules with a wide range of active ingredient contents.[35] Under field conditions these have performed well in long-term delivery of insecticides[36] and herbicides.[37] The sequence of events that occur when the active ingredient is released from this matrix can be visualized using electron micrographs. This is illustrated for a granular formulation containing 45% by weight of the soil and systemic insecticide carbofuran, in Figs. 1-5.[38] Granules with varying degrees of release of carbofuran were obtained by immersion in water, followed by drying, mounting, and coating. The granules were photographed using a Cambridge Stereoscan ZA or a JEOL JSM-T20 electron microscope. The surface of the newly prepared carbofuran-lignin matrix was smooth and carried numerous fine fragments which provided rapid initial release but were not lost following immersion (Fig. 1). After immersion for one day, numerous fissures opened (Fig. 2), which with further active agent loss (21 days

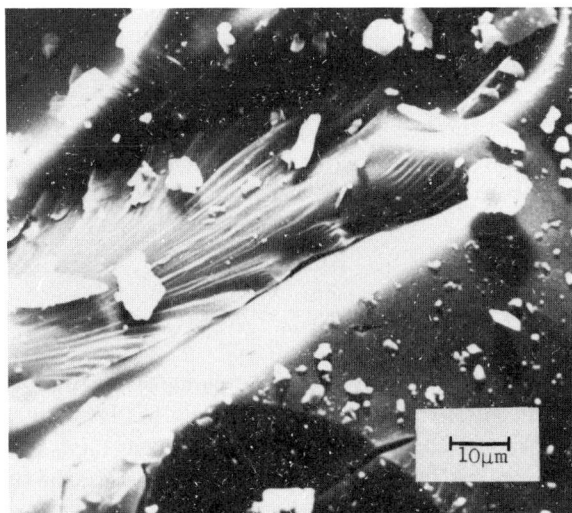

FIGURE 1. The surface of a freshly prepared granule containing 45% carbofuran in a matrix of kraft lignin. Note the presence of numerous small fragments, which may contribute to the high initial release rate.

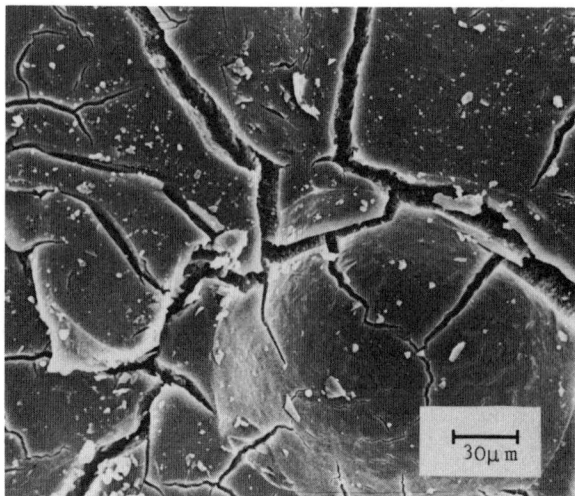

FIGURE 2. The surface of a carbofuran–kraft lignin granule following immersion in water at 34°C for 24 h.

immersion) extended in depth but not in number (Fig. 3). However, even after complete loss of the carbofuran, representing 45% of the structure, the overall dimensions of the granule remained unchanged. This is shown in Fig. 4, which is the exterior of a granule after 63 days immersion or complete active ingredient loss (see Fig. 9). At this stage the interior of the matrix has a completely porous or spongelike appearance as shown in Fig. 5. The integrity of the structure was maintained, albeit in a crumbly nature, in spite of the stresses imposed by such a large proportional mass loss. That the lignin does not revert to its original refractory nature following the loss of the plasticizing insecticide can be explained only by a cross-linking reaction between carbofuran and lignin, but only to a minor degree as the carbofuran can be recovered quantitatively from the immersion water.

2. RELEASE INTO THE VAPOR PHASE

By far the greatest amount of work in the development of formulations for release into the atmosphere or any other vapor phase has involved the release of pheromones to attract or control insect pests.[39] The constraints on these formulations have depended on whether it was intended to release the semiochemical from fairly large and few dispensers acting as baits in traps or from many small sources to cause extensive disruption of insect communication.[40] All these

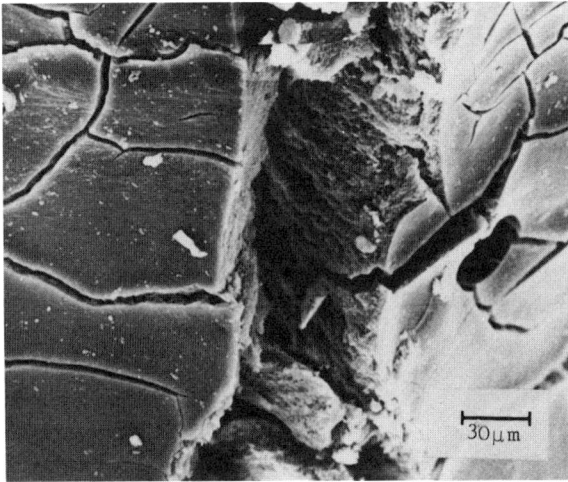

FIGURE 3. The surface of a carbofuran–kraft lignin granule following immersion in water at 34°C for 21 days. Note the distinct appearances of surface and interior structures.

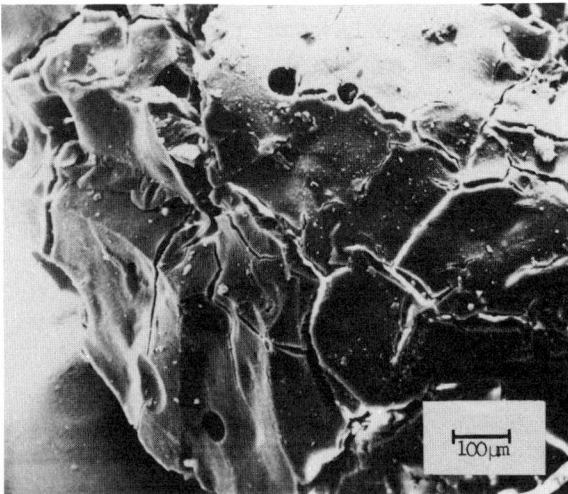

FIGURE 4. A complete carbofuran–kraft lignin granule after immersion for 63 days resulting in complete loss of active ingredient. The overall dimensions are largely unchanged after immersion in nonagitated water.

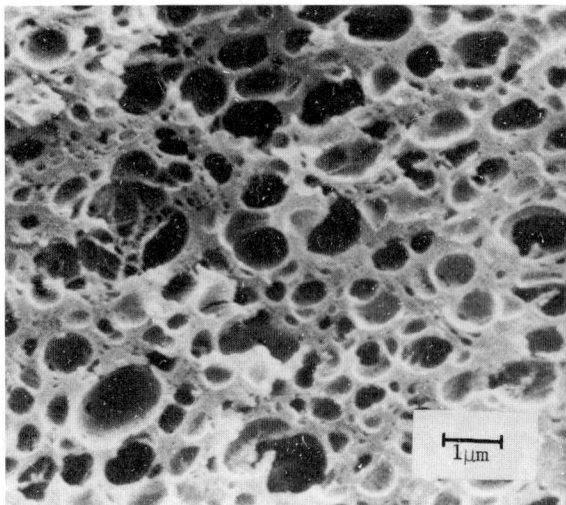

FIGURE 5. The interior structure of a carbofuran–kraft lignin granule following complete loss of active ingredient.

devices release by exclusively physical mechanisms. Although chemical mechanisms of release[41] may be feasible, the structures of the majority of semiochemicals preclude this possibility. Methodology based on the extensive work on insect pheromones is presented in Section 2.1.

Formulations of volatile insecticides were one of the first slow-release pest control applications to appear on the market (e.g., Shell's No-Pest® strip). Again, this represents a physical diffusion controlled-release mechanism. This concept has been applied to various devices such as collars and eartags.[42] Fumigants also function in the vapor phase, and their use could benefit from controlled delivery. Other volatile agrochemicals include the plant growth regulators ethylene and acetylene. Generation of ethylene from ethephon will not be considered as controlled release, as the respective reactions are not independently rate controllable.

2.1 Release Rates of Insect Pheromones

Many formulations have been devised for the controlled release of insect pheromones. Devices used for trapping techniques have included rubber septa, polyethylene tubing, vials, cigarette filters, dental wicks, waxed cardboard, filter paper, PVC pellets, plastic laminates, and hollow fibers. For disruption of mating behavior, formulations based on cork fragments, corn cobs, strings, metal planchets, petri dishes, rubber tubing, microcapsules, polyethylene and polyamide beads, plastic laminates, and hollow fibers have been used.[43] One practical feature of all these formulations is that they lend themselves to release rate characterization under laboratory conditions that simulate their intended location, the atmosphere. This characterization under standard conditions is a very valuable indication of field performance, and depending on the formulation type there can be a good correlation between laboratory and field performance.[44]

2.1.1 Determination under Laboratory Conditions

The rates at which volatile chemicals, including pheromones, are emitted from controlled-release formulations can be measured by three general methods: collection of pheromone after release,[45,46] extraction of pheromone remaining in the formulation,[14,47] and measurement of weight loss.[19] By exploiting the precision in the bore dimensions of hollow-fiber pheromone formulations,[43,48] the change in the length of the liquid column in each fiber can be used in place of weight loss.

Pheromone release rates have been obtained under a range of laboratory conditions by hanging the formulations in a constant temperature room and allowing air circulation and diffusion to transport evaporated pheromone away from the formulation.[14] Increased air movement, such as in ventilated rooms[19] or in a fume hood,[47,49] has been used, but as the rates obtained showed variation, apparatus with controlled air flows has since been developed. In general, measurements of pheromone release were based on weight loss of the formulation.[19,50,51]

Determination of release rates from weight lost by the formulation has two serious drawbacks. If the formulating material contains volatiles or volatile degradation products the release rate will be overestimated. The second drawback occurs when losses of the pheromone arise from degradation which may or may not be reflected in a weight change. A better method is to extract the formulation at intervals of release and analyze for pheromone content. Quantitative extraction of the active ingredient depends largely on the nature of the formulation. Natural rubber septa formulations, much used in insect traps, have been sampled by shaking in hexane-dichloromethane (1:1, 50 ml) with a mechanical shaker for 1 h.[13,14] The resulting solution was then analyzed directly by GLC, using a 2 mm × 2 m column packed with 3% polydimethylsiloxane (SE-30) on 80/100 mesh GasChrom Q, for a range of acetate pheromones and n-alkyl acetates.[14] A similar extraction method is applicable to most pheromone formulations. For example, disparlure content of laminated plastic dispensers was measured[15] before and after a given aging period in the apparatus described below[16] or in the field, by immersion overnight in hexane. Following concentration of this solution, the disparlure content was determined by GLC.

The drawbacks in methods based on residue analysis of the formulations can be overcome only by measuring pheromone actually emitted in the vapor phase. An apparatus for achieving this in the laboratory was first designed by Beroza et al.[16] and provided a laminar flow of air at $32 \pm 1°C$ over nine stainless steel planchets that contained the test materials. The planchets were rotated daily to avoid local variation due to position. The air flow and temperature could be altered according to the requirements of the test and the rate loss anticipated from the test formulation. The rate of release was determined by daily weighing of the planchets or by transferring the sample to a different apparatus and collecting pheromone emitted over a 4-hour period in a constant-temperature environment. This second all-glass apparatus consisted of a petri dish fitted with two air inlet tubes and one outlet tube. The sample was placed in the petri dish, covered, and air dried by passing through a molecular sieve trap passed over the sample at 100 ml/min. The outlet air was then bubbled through solvent (hexane) in a centrifuge tube to trap emitted pheromone. After the period of pheromone collection, the interior surfaces of the petri dish and outlet tube were rinsed with hexane-acetone (1:1) and the washings combined with the contents of the centrifuge tube. The combined solutions were concentrated to 0.1–0.5 ml, and an aliquot (10–20 μl) was analyzed by GLC. With the materials tested [hexadecane and disparlure (*cis*-7,8-epoxy-2-methyl octadecane)], none was recovered from a second trap attached in series behind the first.

The same authors[16] used another apparatus to collect pheromones from cotton wicks. This consisted of a cylindrical glass tube with a joint, terminating in a narrow tube and bent down at right angles. Dry air (100 ml/min) entered through a sintered glass disk, passed over the test formulation, and was led down the narrow tube into hexane (10 ml) contained in a centrifuge tube. As before, the rinsings were combined for GLC analysis. This apparatus has been used in an improved form[15] to allow temperature control by installation in a thermostatically controlled oven. Within the oven, the air flows through a coiled copper tube, permitting thorough temperature equilibration, before entering the glass tube (25 mm inner diameter) containing the formulation under test.

An alternative apparatus has been developed by Cross et al.[52] A wide-mouth Erlenmeyer flask (500 ml), silanized in the interior by using 20% dichlorodimethylsilane in benzene overnight, served as the aeration chamber. A two-holed neoprene rubber stopper carried an air inlet filter (consisting of a 25-ml glass pipet bulb charged with 5 g of 20–40 mesh Amberlite XAD-4 ion-exchange resin preextracted with acetone and then hexane) and a collector device consisting of two glass volumetric pipet bulbs connected by Swagelock unions to each end of a glass tube which passed through the stopper. Collectors were made from 2-, 4-, or 5-ml pipets filled with polymeric adsorbents, normally used in GLC, such as 35–60 mesh Tenax GC or 50–60 mesh Poropak Q. Filtered air was passed over a formulation suspended from the stopper, by means of a vacuum source connected to the collector. Aeration was continued until sufficient (1–50 μg) chemical was collected for reliable measurement with GLC. The stopper assembly was then removed, the flask rinsed with hexane, and the collectors eluted with 20% ether in hexane. The total solvents (7 ml) were concentrated under nitrogen and analyzed. Recovery was quantitative, and little oxidation occurred with the chemicals used. After washing and drying, the collectors and adsorbents can be reused many times. A variety of other adsorbents have been used to collect pheromones including charcoal, lanolin, triglycerides, and glass surfaces.[49]

In an effort to reduce the time required for each pheromone assay when determining release rates from controlled-release formulations, thermal desorption from the collectors has been suggested.[45] The pheromone dispenser was suspended in a vertical glass tube, and air was drawn through a charcoal filter over the dispenser and through the collector containing 150 mg of Tenax GC adsorbent. The overall length of the apparatus was 14 cm. Using silanized glassware, a high airflow rate (100 ml/min), and close proximity of the dispenser to the collector, a high proprotion (88–95%) of released pheromone was adsorbed onto the collector. After aeration was completed the glass collector tube was transferred to the heated injector block of a GLC apparatus with the adsorbent located within the injector and the distal end of the tube connected to the column. Desorption (90% efficiency) required 4 min (at 260°). Analysis by GLC thus follows immediately on aeration, and two or three release rate determinations can be performed daily. Further developments with this approach are likely to provide a degree of automation.

In a contrasting approach, Baker, Cardé, and Miller[49] have collected pheromone from controlled-release dispensers and from live moths under static air conditions. A sealed round-bottom flask (250 ml) provided both the container and the collecting surface. The test object was suspended in the flask and after a time the flask was opened, the dispenser or insect was removed, and the flask was washed with small quantities of redistilled hexane. The hexane solution was concentrated and the pheromone content estimated by GLC using an internal standard. Collection efficiency by this method was 100% even at nanogram levels. Careful methodology is required to reduce extraneous substances and maintain sensitivity in analysis. This method is appropriate for measuring release rates of pheromones from different substrates without air flow, such as those from insects that release under still-air conditions. The importance to release rates of a diffusion boundary layer could also be evaluated by comparison with the previously described air flow measurements.

The problems involved in release rate determinations of volatile and often unstable chemicals have been emphasized by Weatherston et al.[43] When the same formulation has been studied by different techniques, different release rates have been obtained. Release from hollow fibers containing gossyplure was 1.50–2.86 μg fiber per day (measuring pheromone remaining in the fibers), 0.04 μg per fiber per day (vapor-phase content from static air and air flow methods analyzed by GLC), and 0.34–0.74 μg per fiber per day (from both effluent collection methods estimated by scintillation counting). These authors stress the importance of reliable methodology in enabling the study of the effects of climate, degradation, formulation, and other factors on release rates.

2.1.2. Determination under Field Conditions

The rates of release of pheromones and other semiochemicals from controlled-release formulations depend on temperature,[46] air movement, and possibly humidity.[53] Thus measurements made under laboratory conditions, where these variables are taken into consideration, will give some indication of release rates in the field. However, one important feature of the field is the variation of these factors over time and space. Often, the variation in release rates, such as that due to temperature, works unfavorably for optimum control of insect mating.[54]

An additional complication in the determination of release rates under field conditions is degradation of the active ingredients. This has considerable importance in laboratory assays[55] with certain chemicals and is likely to be of greater importance in the field.[54] The occurrence of this problem can be ascertained only by carrying out analysis and release studies on the same formulation.[54,55]

Where large dispensers are used in field-located trapping and monitoring systems, the pheromone release rate may be studied by extraction and analysis of retrieved sample formulations. This method has been applied to a range of dispenser types, including rubber septa,[13] plastic laminates,[46] and polyethylene tubing used for mating disruption.[56] Analytical methodology is similar to that discussed in Section 2.1.1 for laboratory studies. Sprayable formulations (e.g., microcapsules) can be applied to artificial surfaces, held in various parts of the treatment area, and retrieved for analysis.[54]

By far the most usual indirect method of determining release rates and the efficacy of the formulation is with bioassays. As this method is complicated by the possibility of a mixture of chemicals in the natural pheromone and the uncertain nature of the response in any population of the insects used for the bioassay, it can give only an indication of release rates and usually only response threshold values. Bioassays using cultured insects would seem to be preferable for reproducibility of results[57] but may differ from field populations. Bioassays can also be used under laboratory or greenhouse conditions. Greenhouse bioassays using cultured california red scale, *Aonidiella aurantii* (Maskell) were used to follow the release of (3Z, 6R)-3-methyl-6-isopropenyl-3,9-decadien-1-yl acetate from rubber septa.[58] The bioassays were conducted by measuring the relative attractancy of formulations (following field weathering in the sun, in shade, and at elevated temperatures) compared to virgin females placed in traps on a turntable surrounded by mature male and female red scale insects. The release rate was estimated as a percentage of the total male capture (which never exceeded 40%). Thus, this bioassay is useful when the release rate falls to the threshold value for response by the males. It also indicated that the synthetic pheromone is stable at temperatures likely to be encountered in field use conditions.

Field bioassays based on attractancy can be considered synonymous with the practical use of pheromones as baits in traps. However, in terms of evaluating the rate of release from controlled-release formulations, the results thus obtained need to be considered in the light of knowledge of the behavior and populations of the pest insect. At best, without detailed measurements on the total population, the trapping frequency is comparative.[40,59]

Where the formulation is intended for air permeation with subsequent mating disruption, release rates are more difficult to quantify in the field. A number of different bioassays and assessments can be carried out in the field to determine the presence of pheromone at time intervals following application. These techniques include counting insects attracted to traps placed in the area permeated with the pheromone, baited with dispensers releasing the same pheromone[60] or with calling females,[56] and comparing the numbers with insects caught in a control area. Other workers have used mating tables carrying tethered females[61] or cages containing male and female insects[62] and have assessed the extent of mating by dissection or counts of fertile eggs laid. Information on pheromone release in time and space can also be obtained through the use of release-capture studies of dye-marked insects (e.g., screw worms[63]). For discussions of methods used in evaluating the success of mating disruption such as depression of subsequent larval populations and damage levels, the reader is referred to the many accounts of field trials.[40,63,64]

The effects of environmental and climatic factors on the release and dissipation of pheromones from controlled-release formulations has been studied in a few situations. Quantitative information on pheromone in the atmosphere shows that there is considerable variation according to the environment and the location of sampling. This information is normally not available from biological studies using the target insect. Release from gelatin microcapsules of disparlure applied to a grass field was followed for two months by air sampling.[44] Two formulations (2 and 10% disparlure dissolved in 3:1 xylene-amyl acetate) were sprayed at 500 g active ingredient per hectare. Samples were taken by drawing a known volume of air through a sampler (a glass tube 3.5 cm inner diameter, 20 cm long) with a vacuum pump ($2-3 \text{ m}^3/\text{h}$) and entrapping the pheromone on a molecular sieve. Samples were taken at five different heights, depending on the growth of the grass, and up to 190 cm. After sampling, the molecular sieve was extracted with 1:1 ethyl acetate-hexane (10 ml), concentrated to 0.2 ml, and treated with 0.25 M triphenyldibromophosphorane in dichloromethane (0.5 ml). The resulting dibromo derivative of disparlure was cleaned up on Florisil and analyzed using GLC with an electron-capture detector. The limit of sensitivity was 0.2 ng disparlure/m^3 air. Wind speed and air temperature were recorded for each sampling height, and the chemical flux rate was determined. Concentrations of disparlure decreased with height, and vapor flux rates were found to depend strongly on temperature and also on the formulation type, with the 10% micro-

capsules giving better release characteristics for controlling the gypsy moth
[*Lymantria dispar* (L.)].

A similar approach has been taken to study the release of (Z)-9-tetradecen-1-
yl formate, a mating disruptant of two *Heliothis* species, from laminated plastic
dispensers (97 cm² in area, containing 300 mg pheromone and releasing approximately 75 µg/h) spaced 10 m apart in a corn crop.[53] Air concentrations decreased
with time and 27 days after treatment were 0–30% of those on the sixth day.
The corn plants and the soil appeared to act as sinks for the pheromone vapor
at night.

Release rates determined under field conditions may be substantially different
from laboratory-measured rates as a result of weathering.[65] Exposure to sun and
rain reduced the time for 50% loss of disparlure from microcapsules from 123
days in the laboratory to 10–34 days.[61]

2.2. Other Vapor-Acting Semiochemicals

In contrast to the extensive studies on the use of pheromones, the development
of other semiochemicals for crop protection use is very limited. For those substances acting in the vapor phase, controlled-release formulations will be needed
for effective application. Behavior patterns elicited in insects and other pests can
include attractancy, repellency, antifeedant, oviposition-deterring, and others.[40]
The advantages of using a substance that inhibits pest behavior lies in the potential
for use in a manner similar to conventional crop protection agents; it can be
applied by the growers themselves to their own crops and without cooperation
with other organizations. These chemicals, especially those with a broad spectrum
of activity, could have a promising future in crop protection.

Attractants, such as those originating in the pest's host plant, can function
similarly to sex attractants, as baits in traps. In this manner, allyl isothiocyanate
has been used[66] to trap flea beetles when released from glass vials sealed with a
rubber stopper to which the liquid attractant was delivered by capillary attraction up a pipestem cleaner. The rate of release through the stopper was measured
by weight loss of vials exposed at different levels below a 500-W floodlight in a
fume hood with a constant air flow for 14 weeks. These measurements indicated
a gradually increasing loss rate (from 200 mg/week) over 8–9 weeks (to 260 mg/
week) and subsequent decrease to 220 mg/week at the 14th week. In the field
these vials attracted a proportion of the flea beetle population, but as no dosage-
response data were taken no estimate of the efficacy can be made.

Repellents that act in the vapor phase must be dispensed by controlled-release
formulations to be used effectively in crop protection programs. In an attempt
to obtain controlled release of naphthalene to repel ovipositing cabbage root fly
(*Delia brassicae*), various formulations were applied to the soil around the stem
base of Brassica plants in the field.[67] Release rates were assessed indirectly by

counting pupae and scoring root damage. Only moderate levels of protection were achieved. This result underlines the importance of evaluating release rates and selecting appropriate formulations before attempting field trials.

Slow release of a repellent has been claimed in another insect control application. The repellent, 3-tolyl N-methyl carbamate, was incorporated into a solid composition of adamantane and cyclodecane.[68] When suspended in peach orchards, this formulation reduced the number of scars per peach caused by ovipositing moths from 24 in the control area to 2 during a total period of 16 days.

2.3. Release Rates of Fumigants and other Volatile Pesticides

Fumigants and other volatile pesticides are often released into the air phase of enclosed spaces. For very volatile and toxic materials (e.g., methyl bromide) the enclosed space is usually sealed. In these cases, the rate of release of the fumigant is less critical than in the open, and the important parameter is the total amount released. Release rates can thus be estimated from the concentration of the fumigant obtained by sequential air sampling of the enclosed space.

Phosphine for fumigating stored products is generated from metal phosphides by the action of atmospheric moisture.[69] The release of phosphine can be delayed by reducing the access of moisture, using a fibrous cover.[70] The rate of release of this type of formulation in the shape of a 7-mm thick plate (containing magnesium phosphide) at 28°C and 60% relative humidity has been studied by sampling the accumulated phosphine in a fumigation chamber. Maximum phosphine concentration occurred in 32 h; doubling the area of the plate increased the time to maximum concentration to 56 h. Almost total release was achieved in these periods. Air in the fumigation chamber was sampled and removed through a tube, and the phosphine was estimated colorimetrically.[71] As a consequence of the variability of response of different insect stages to phosphine and the importance of the duration of exposure,[72] bioassays with insects are less practical for determining release rates.

Formulations for controlled release of ethylene dibromide and p-dichlorobenzene in polyethylene film packets have been developed to control the greater wax moth in honeycombs.[73] Rates of release were determined over a range of temperatures (10–35°C) by weight loss when dispensers were hung in a controlled environment chamber. In a separate test, vapors emanating from the formulation containing the mixture of fumigants were recovered in ice-packed bubble traps containing redistilled hexane. Aliquots taken from the traps were examined using GLC.

Release of volatile insecticides from plasticized polyvinyl chloride have usually been measured by weight loss. A PVC strip containing dichlorvos lost 6% of its weight during its effective lifetime of 16 weeks.[74] As this formulation contains possibly 30% of a plasticizer and both this and dichlorvos are emitted, this is not a very accurate method.[74] As the release is diffusion controlled, the rate of release falls with the square root of time.[5] Efficiency for house fly control can be improved by utilizing a nonvolatile insecticide combined with a volatile attractant. This is the principle of a three-layer plastic laminate system.[75] Release of the house fly food attractant, vanillin, was followed by weight loss. A linear release rate was achieved over the effective life of 3 months, but only 20% of the initial vanillin loading was emitted.[75]

Other systems have been devised for the release of volatile materials, such as deodorants, perfumes, fumigants and insecticides. These systems include gels and a matrix of polyethylene paraffin and a granulated, partially dehydrated metal salt.[76] In general, the release rates have been characterized by weight loss only.

3. RELEASE INTO AQUATIC ENVIRONMENTS

In many respects, methods for the determination of release rates under aquatic conditions have fewer problems than for determination in the vapor phase. Formulations for releasing pesticides and behavior-modifying chemicals can include both physical and chemical systems. Physical systems used in aquatic environments are usually of the matrix type, releasing the active ingredient by a diffusion-dissolution or leaching process. In chemical systems the aquatic pesticide is covalently bound to a polymer, and release occurs through hydrolysis and biological degradation. Methods of assessment of release rates for the two different systems will differ, especially in the case of laboratory tests. Target organisms of particular importance include weeds, snails, and insects, but this section will be concerned with methods for the determination of release rates of aquatic herbicides, as their relevance to agriculture is greater.

3.1. Release Rates of Aquatic Herbicides

The early development of controlled-release aquatic herbicides has been described by Cardarelli.[77] It can be seen that the physical method of obtaining controlled release has been the most successful, and indeed a pelleted natural rubber matrix containing 2,4-D butoxyethanol ester has reached commercial standing.[78] The use of an elastomer as a matrix has advantages in the range of aquatic herbicides that are compatible with it and its ability to be processed in a multitude of forms and densities to provide floating and sinking applications.[79] Release mechanisms involve diffusion-dissolution and leaching processes.

During the development of these formulations, experimental methods for determining release rates have been employed. Generally, small containers or aquaria under controlled conditions in the laboratory are used initially.[79] Analysis of the water at intervals following immersion can be carried out spectrophotometrically or by atomic absorption spectrometry.[80] Release rates measured using this simple system have shown dependence on the diffusion coefficient of the herbicide in the water, the initial concentration of the herbicide, and the formulation geometry.

More frequently, evaluation of controlled-release herbicides has been based on bioassays, and to anticipate possible environmental hazards a range of organisms in addition to weeds has been used.[81] For aquatic weed bioassays the choice of plant species is important and should, of course, include anticipated target organisms. Ideally, the dosage–response relationship should be known. Assessments are based on percent mortality compared to the control as determined by observation of browning and disintegration.[79] The plants used in these bioassays[80] have included *Myriophyllum spicatum, Vallisneria americana, Cabomba caroliniana,* and *Elodea canadensis.* Larger tanks and outdoor pools have been used as an intermediate stage between laboratory tests and field trials. To evaluate the concept of herbicidal activity localized in the region of the target weed (the phytozone) a 6-ft tall transparent plastic tube was filled with distilled water.[79,82] Candidate formulations were introduced at different vertical positions, and subsequently aliquots of water were withdrawn through taps positioned along the length of the tube. Water analysis indicated that a slight upward movement of the herbicide (2,4-D butoxyethanol ester) occurred. Thus the concept of phytozone treatment was vindicated.

Although field trials have been conducted with herbicides in elastomer formulations,[79,75] no measurements of release rates under those conditions have been published. Due to the uptake of aquatic herbicides by mud and organic matter, the rate of release under field conditions could be measured conveniently only by retrieving the formulations and analyzing for residual herbicide. However, reports of field tests have not referred to this line of investigation, and thus the evidence to support the claims for release lifetimes based on laboratory test data[77] have not yet been substantiated.

For chemically bonded controlled-release aquatic herbicides, release rates will depend upon the environment in which they are placed, as the required degradation will depend to a large extent upon microbial activity. Experimental studies of release rates for chemically bonded systems have been based upon chemical hydrolysis in sterile buffered aqueous solutions maintained at a constant temperature.[83] Here a pH of 8 was used to simulate natural waters of the American south. The amount of herbicide released was determined at intervals by spectrophotometric analysis of the buffered water. After 296 days of immersion of a 2-acryloyloxyethyl-2,4-dichlorophenoxyacetate–methacrylic acid (90:10) copolymer, 87% of the 2,4-D originally contained in the polymer was recovered from the aqueous phase. The rate of release under these conditions increased during the first few days and then remained relatively constant. As the expected first-order degradation kinetics were not followed, it was assumed that the rate constant for the hydrolysis reaction continually increased as the 2,4-D content of the polymer fell.

Bioassays carried out under laboratory or field conditions rarely yield release rates. Instead, information is provided as to whether (1) a candidate formulation actually releases the herbicide and (2) the duration of release occurs at a level sufficient to cause observable effects. Laboratory bioassays have been used to evaluate 2,4-D bonded to polystyrene by ester and anhydride linkages and provided information on release of herbicide for 7 and 14 days after initial immersion.[84] Percentage inhibition of growth of *Lemna minor* was estimated and compared to the control.

4. RELEASE INTO SOIL AND PLANT ENVIRONMENTS

In agricultural and forestry applications, release on and into soil is the most important compared to the other areas already considered. For convenience, we shall include here release to plant surfaces, since the variation in climatic and biological conditions is similar to that in soils. A characteristic of soil is its complexity, and depending on the nature of the active ingredient, movement in soil to the target organism can occur through diffusion or mass flow in either the vapor or aqueous phase or in both.[8] Of particular importance to transport is the adsorptive capacity of soil, especially the clay and organic matter fractions. In many respects, soil itself functions as a slow release substrate, and when choosing candidate pesticides this possibility must be considered. The desirability of formulations releasing by zero-order kinetics is therefore not as paramount in soil as in other situations, as the dosage-time profile as received by the target will be modified by adsorption.[8] Selection of soil-applied pesticides for controlled delivery should therefore be based partly on a consideration of their adsorption and mobility in soils.

Thus, in assessment of release rates, the correct experimental design is important. Initial release studies under abiotic laboratory conditions are useful for rapid comparison of candidate formulations and for determination of fundamental properties, such as diffusion constants, necessary for rational design of formulations.[85] However, it must be recognized that release mechanisms observed in such tests may not be rate controlling under field conditions. Even with physical release systems where diffusion through the polymer matrix is normally rate controlling, under soil conditions where moisture levels could be insufficient, removal of the active ingredient from the surface of the formulation to the surrounding environment could be rate limiting. This effect has been elegantly demonstrated by alternately wetting and drying formulations under artificial conditions.[86] Other types of behavior may also be important, depending on the formulation and the physical properties of the active ingredient.

With chemically bonded controlled-release formulations, the choice of assay is more important. In this case, and also for release by matrix erosion, the rate will depend on environmental factors such as moisture levels, soil nutrition, temperature, and microbial activity. Hence, release rates, as measured in sterile hydrolysis studies may have little in common with kinetics obtained in the field. Also, under field conditions degradation of the released active ingredient is taken into account, whereas in water immersion release studies there is normally very little breakdown. Laboratory release studies can be very valuable once a correlation between these and field studies can be built up from experience. Release rates determined in soil under controlled conditions provide a useful intermediate experimental assay most often employed for chemically bonded pesticides.[22,29,87] Again, there is a disparity compared to field conditions. Release in the field may be faster or slower[3] depending on the relative conditions of the two environments. In soil, analytical problems are more difficult, and for a complete picture of release rates and degradation, recourse would have to be made to the use of radiolabeled compounds. Thus, bioassays are frequently employed and again will normally only reveal whether sufficient release occurs and for how long.[22]

4.1. Release Rates from Physically Bound Formulations

Controlled-release formulations for soil application that contain the active ingredient in a physically bound form include encapsulation of liquids and solids and matrix systems that release by diffusion, leaching, and erosion. Geometry is an essential feature of physical systems. Where good distribution in soil is a prerequisite for biological efficacy, there is a limitation on the dimensions of the individual particles. This sets practical limits on the performance of the formulation with the inevitable tendency to fast initial release followed by a declining release rate. Retrieval and analysis of individual particles is proscribed by practical considerations in the field studies, and thus both water immersion tests conducted under defined abiotic conditions and vapor-loss methods provide quantitative data on release characteristics of physically bound formulations.

4.1.1. Determination under Laboratory Conditions

Release rates can be measured by immersion of the sample, in the form of granules or larger pellets, in water under laboratory conditions. The larger sized formulation with a known surface area can be used for the evaluation of the parameters of release. The release rates of hexamethylphosphoric triamide, a house fly chemosterilant, from a matrix of polyethylenedilinoleamide were determined by immersion of a block of the pesticide–polymer combination in stirred distilled water.[85] Samples of the liquid were periodically withdrawn and the active ingredient was measured from the phosphorus content determined by neutron activation analysis. The release rates were found to depend on the square root of time, in agreement with diffusion theory.[85] Determination of the diffusion coefficient and the nonideality coefficient in this way allows for more rational selection of polymers for controlled-release formulations. In a similar method for studying release rates, a diffusion-controlled mechanism was established for the release of the herbicide carbetamide from a matrix of kraft lignin.[88] By plotting the released herbicide per unit surface area against the square root of time, a straight-line relationship is obtained (Fig. 6) for each of the active ingredient contents tested. When expressed in terms of the proportion of active ingredient released, a constant slope is observed (Fig. 7).

The immersion test may be modified in an attempt to simulate the anticipated field environment of the test formulation. Thus, the water can be maintained in static,[89,90] shaken,[20] or stirred[85] conditions and small samples removed periodically, making up to constant volume with fresh water, or the entire volume of water can be changed at each measurement. The results obtained with each approach can differ and may not be strictly comparable. When the released active ingredient is allowed to accumulate in the immersion liquid, dissolution from the matrix surface may be reduced. The volume of water used should be of sufficient solvent capacity not to be exceeded by the amount of active ingredient released

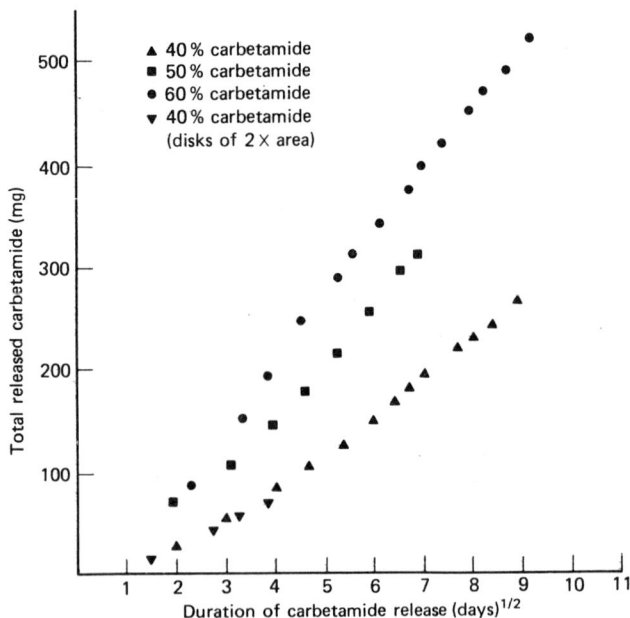

FIGURE 6. Results from a static water immersion test: release of carbetamide from a kraft lignin matrix per unit area of a disk-shaped formulation.

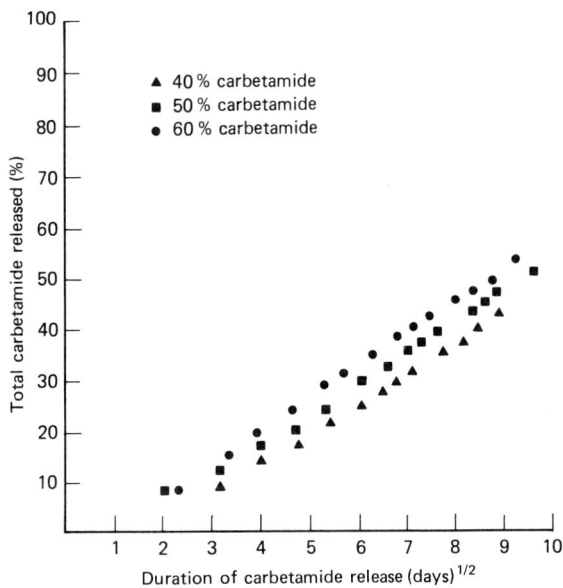

FIGURE 7. Release rates of carbetamide from a kraft lignin matrix. This is the data of Fig. 6 replotted to give release rates as proportions of total available active ingredient.

into it. Under static water conditions the diffusion boundary layer may exert greater influence on the rate of release, depending on the polymer and active agent.[6]

In attempting to simulate the effects of a soil environment with varying moisture levels, intermittent immersions have been used. For example, the rate of release of the herbicide alachlor from plaster-of-paris tablets was assessed by immersion in water for 15 min daily over a period of 8 weeks.[91] After immersion the water was removed by filtration under suction and analyzed using GLC apparatus fitted with an electron-capture detector.

The water for the immersion test is normally sampled and analyzed by spectroscopic, chromatographic, or colorimetric methods.[29,92] This approach is particularly convenient as there are usually few contaminants to interfere with the assay. Another method, already referred to, is to follow weight loss. This is appropriate where the formation is large enough to be handled and where there is no breakdown of the matrix. Large (1 cm diameter) cylindrical pellets of a fatty polyamide containing 25% insecticide have been successful in controlling the mahogany shootborer for up to one year.[10] The best performance was provided by the matrix containing carbofuran, and the release rate of this was assessed by weight loss into water.[93] The pellets were held in vertical glass tubes with the bottom ends closed by taps. Water was added weekly, allowed to remain for

24 h, and then drained away. At other times the pellets were kept humid and at a constant temperature (24°C). Release rates were determined by spectrophotometry of the carbofuran in the rinse water or by removing the pellets, air-drying them under standard conditions, and weighing them. This testing method was designed to assay formulations intended to release systemic insecticides for long-term control of the mahogany shootborer by placing the pellets in the soil root zone of young *Cedrela odorata* trees growing in the humid tropics of Central America. The release characteristics obtained in this way are depicted in Fig. 8 and are much as anticipated from diffusion theory,[85] except for the maximum occurring about 90 days after testing started. This was most likely due to a surface effect generated during the casting of the melted plastic combination.

The concept of intermittent immersion in water can be used to further extract information on release mechanisms. This approach was employed by Bittner and Perry[86] when they studied the release of the synthetic cytokinin benzyladenine from a copolymer matrix of methacrylic acid and methyl methacrylate (1:1). Two distinct release rates were observed: one initial rate governed by diffusion through the surface film and (after 3.5 h immersion) a subsequent rate controlled by the slower diffusion within the polymer particle. This hypothesis was tested by removing the polymer particle from the water at 7.5 h, drying it, and storing it at room temperature for 24 h. During the dry period, internal diffusion continued until equilibrium was established, and then when the particles were re-wetted release was resumed at the initial film-controlled rate.

Controlled-release formulations may also be evaluated by exposure to flowing water.[94] As a consequence, diffusion boundary layer effects are reduced, and, especially for porous matrices, accelerated release is observed. This method has advantages in that it can be made semiautomatic. For example, release rates of salts of dicamba and picloram and esters of 2,4-D have been measured in this way.[90] The candidate formulation was loaded into a bed of quartz sand in a glass column and distilled water delivered from above at the rate of 2 ml/min. The column was mounted on an automatic fraction collector, and selected fractions were bioassayed for herbicide content using a cucumber root length assay. In general, for the flowing water test the exterior layers of the matrix are exhausted more rapidly, giving a longer subsequent low release rate compared to static immersion tests. This can be illustrated by considering the release rate of carbofuran from a kraft lignin matrix containing 45% active ingredient. When determined by a static water immersion test, the release curves are shown in Fig. 9. Under a flowing water test (14 ml/h), release from the same granular formulation has the characteristics depicted in Fig. 10 by curve A.[95] However, the value of this test is in comparison of formulation performances. Thus it can be seen from Fig. 10 that the lignin granules have a release rate distinct from another type of carbofuran slow-release formulation in which a clay-based granule is coated with a polymer film (curve B).

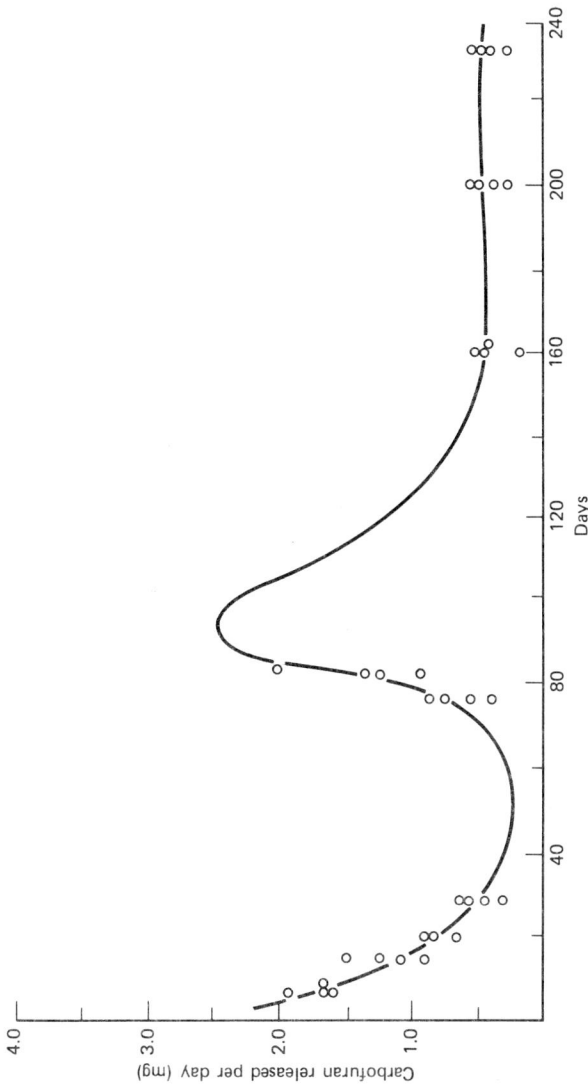

FIGURE 8. The rate of release of carbofuran from a polyamide matrix determined by weight loss measurements.

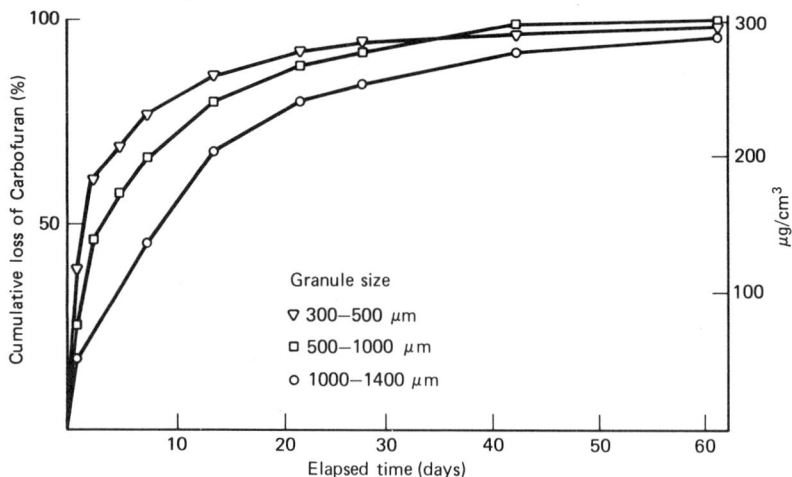

FIGURE 9. The rates of release of carbofuran from a kraft lignin granular matrix containing 45% active ingredient.

For many pesticides, movement in soils and uptake by organisms can occur in the vapor phase.[8,96] Although beneficial in maintaining good distribution in dry soil, these pesticides can also suffer losses to the atmosphere after application. As a result of competition for adsorption sites on the soil particles, losses are greater when the soil is wet. In practice these losses are reduced by incorporation into the soil. Controlled-release formulations of these relatively volatile chemicals can be assessed by exposure to a continuous flow of air. A range of starch-encapsulated formulations has been tested in this way. In a replicated series of tests, starch xanthide granules containing a nematocide, 1,2-dibromo-3-chloropropane (DBCP), were exposed to constant aeration in a forced-draft fume hood at 22° for 4-day periods. Release of the volatile nematocide was determined by weight loss or by total halogen content.[97] In a previous report[98] these workers evaluated DBCP and diazinon starch granules by aeration, wet or dry, in open dishes for periods varying from 10 to 44 days in a fume hood. Assays were made using the saprophytic nematode *Panagrellus redivivus* and by weight loss. Comparisons of release rates were made with technical grade DBCP and diazinon and with the encapsulated formulations kept in sealed vials. The rates of release were much faster when the starch granules were wetted. Nematocides in microcapsules have also been evaluated by this technique.

A less severe test of formulations of volatile soil pesticides is to avoid forced ventilation. Starch-encapsulated herbicides (EPTC and butylate) have been evaluated by standing them in open flasks for 17 days.[99] This report also included an evaluation based on a water immersion test; thus both tests represented the two transport phases of the soil medium.

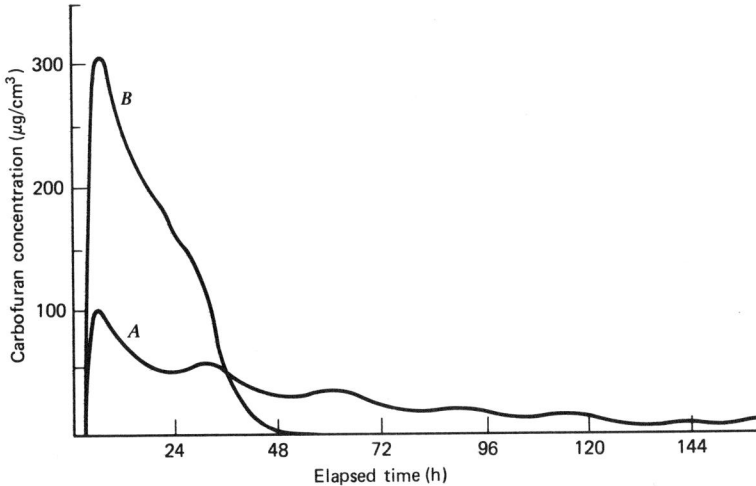

FIGURE 10. Release rates determined under flowing water conditions comparing (*A*) carbofuran–kraft lignin granules (45%) and (*B*) plastic-coated clay granules containing 5% carbofuran. Flow rate = 14.4 cm³/h; temperature = 15–20°C.

Release rates from microcapsules have been determined by immersion in aqueous buffer solution using a rotating bottle apparatus[100] and analyzing the water for the active ingredient (malathion). Microcapsules, as manufactured, usually have a wide range of physical properties, differing, for example, in wall thickness, diameter, and uniformity. These factors can have a profound effect on release rates. The variation between individual microcapsules has been studied by Dappert and Thies,[101] who found considerable variation in permeability to hydride ions of a wall composed of partially hydrolyzed ethylene-vinyl acetate copolymer cross-linked with an isocyanate adduct.

4.1.2. Evaluation under Soil Conditions

The evaluation of controlled-release formulations in a soil medium in the laboratory or greenhouse using columns or pots is a useful intermediate stage between abiotic tests (such as water immersion) and field trials. However, the problems of analyzing the released or bound pesticides implies too great an investment of effort, and most workers have employed bioassays to monitor release rates. In addition, it is difficult to extract the active agent from the soil without liberating the pesticide that still remains bound within the formulation granules. However, with water-soluble agents, continuous extraction and measurement is possible without undue perturbation of the formulation.

This has indeed been the approach with controlled-release fertilizer formulations and is a standard method in fertilizer studies. To give one example, the release of urea from sulfur-coated formulations was obtained by mixing with soil in oxygen-permeable polythylene bags for 1, 3, and 6 weeks.[102] Following incubation, the soil was transferred to beakers, distilled water (500 ml) was added, the mixture was stirred well, and the pH measured. Nitrate was determined by a specific ion electrode and nitrite by colorimetry. Potassium chloride was added, and the mixture was stirred and filtered. The filtrate was analyzed for urea by 2,3-butanedione monoxime and thiosemicarbazide complex at 527 nm and for ammonium ion by Nesslerization at 410 nm. To complete the nitrogen balance, the soil was wet-sieved through 2-mm and 1-mm sieve openings, the controlled-release pellets were removed by tweezers and crushed, and their nitrogen content was determined by the macrokjeldahl method.

A simplified method of continual determination of nutrient release has been suggested by Holcomb.[103] In place of soil columns, plant pots containing the growing medium and the controlled-release formulation were used. The medium was watered to capacity and the pots placed on capillary matting inside a closed container. The mats were removed at intervals and the solution in the mat extracted by squeezing. The mat was returned to the container and resaturated with distilled water, and the pot was then replaced on the mat. The potassium content of the extracted solution was analyzed using a specific ion electrode. This method was used to follow the potassium release rate of several membrane-coated fertilizer formulations.

Quantification of release rates under soil conditions can be achieved by retrieval of the formulations at varying intervals. Detailed studies reported have used controlled-release granules sealed in Saran screen packets to permit recovery at specified times after treatment.[20]

Bioassays can provide a certain amount of information regarding release rates in soil. Two approaches are feasible with small-scale pot tests: (1) to extract the released agent from the soil and perform a bioassay to quantify concentration and (2) to obtain semiquantitative results using organisms placed *in situ*. The second approach is more convenient and results in fewer errors, since the extraction stage is eliminated. However, when this method is used, the visual symptoms of the biological response are delayed and no specific time interval can be ascribed to the measurement.

4.2. Release Rates from Chemically Bound Formulations

Formulations containing the active ingredient chemically attached to a polymeric substrate are normally intended for environments that contain the agents that break the attachment. For soil and plant applications the dominant mechanism is microbial breakdown.[3] Hence, release rates determined under abiotic conditions cannot possibly reflect the situation in practice but provide useful comparative data for design purposes.

4.2.1. Determination under Laboratory Conditions

Hydrolysis studies are usually carried out by suspending the formulation in water buffered to certain pH values. This approach has been developed in the series of reports discussing the attachment and release of the herbicide metribuzin to a variety of polymeric substrates.[21,83,104,105] The polymers were placed in an Erlenmeyer flask with water (500 ml) and stirred. At designated intervals the water was sampled and analyzed. Metribuzin contests of the water samples were determined by ultraviolet spectroscopy, by extraction into benzene with subsequent electron-capture GLC, or by liquid chromatography. Later studies employed smaller amounts of polymer (0.05 g) in distilled water (20 ml) and smaller sample volumes (10 μl). Release was determined over a period of 80 h.

Other workers have examined the effect of pH on the hydrolysis kinetics with a view to simulating soil conditions. Mehltretter et al.[27] theorized that, in the case of starch esterified by 2,4D-,acid soils would hydrolyze the glucoside linkages in the starch, while alkaline conditions would saponify the ester bond. Thus the hydrolysis studies adopted by these workers used phosphate buffer solution at pH 6 or 8. The controlled-release formulation was added, the container closed, and the contents stirred with a magnetic stirring bar. For assay, the mixture was centrifuged, the sample removed, and an equivalent volume of fresh buffer solution replaced. Release rates were measured for 100 days, and the results supported the above hypothesis regarding hydrolysis mechanisms.

In studying the release from the related 2,4-D esters of α-cellulose, the polymer was enveloped in four layers of porous teabag paper.[106] Bags containing 125 mg of active ingredient were suspended in unstirred distilled water (500 ml) adjusted to pH values of 4, 7, and 10 using sodium hydroxide in hydrochloric acid and maintained at 22°C. For assay the bags were transferred to fresh solutions, while the residual solutions were analyzed for 2,4-D content using a Dohrmann Environtech DC-50 total carbon analyzer. Release rates were found to follow a first-order relationship but with two distinct slopes. These two different rate constants were explained on the basis of two different molecular environments of the ester bonds, depending on the crystallinity of the cellulose.

Similar first-order kinetics were observed with starch and cellulose to which benzyladenine and other cytokinins were attached by carbamate bridging structures.[107] These polymers were suspended in water at pH 1.5 and 12, and the cytokinin content of the water was measured by absorption at 260 nm. The release rate was measured for 600 h, at which time total release was obtained. Bioassays were also conducted on the water, using soybean callus growth.

4.2.2. Determination under Soil Conditions

It has been emphasized already that hydrolysis studies *in vitro* may not correlate with chemical breakdown of formulations in the field. The use of pot experiments with live soil, of cultures containing one or more microbial organisms, or of enzymes can all provide an intermediate stage of testing the breakdown of polymers. In the last two approaches, organisms or enzymes appropriate to the intended applications need to be selected. Very few studies of this type have been reported. One example would be the characterization of starch-encapsulated pesticides by measuring the amount of free glucose released under standard conditions by treatment with diazyme.[108] Although it is not a chemically linked type of formulation, release from this type of matrix depends on the degradation of the starch xanthide.

Apart from short-term availability studies[109,110] in soil under laboratory conditions in which the released pesticide was extracted and measured by GLC, there have been few reports on quantification of release rates in soil environments. One study refers to the use of a mass spectrometer to quantify 2,4-D release from an ester-bonded formulation.[111] The "residual phytotoxicity" for polymeric derivatives of metribuzin has been measured in a controlled environment chamber using a mixture of weeds susceptible to the herbicide[109] or soybeans and the weed prickly sida (*Sida spinosa* L.).[108] Plants were allowed to grow for 14 or 21 days and the bioassay was repeated at intervals up to 112–120 days. With herbicides that inhibit seed germination the quantal response can be evaluated more rapidly, and with lettuce 50% germination occurs with 2 ppm of 2,4-D. As germination can be observed in 2–3 days, this response provides a convenient, rapid method of *in situ* soil bioassay.[22] An equally sensitive method for 2,4-D is the root elongation test of cucumber seeds conducted under controlled conditions.[106]

In a study or a range of polymeric derivatives of 2,4-D (see Table 1) under soil conditions outdoors, a series of pots were set up containing varying amounts of the polymers.[22] Lettuce seed bioassays were continued until inhibition of germination ceased. At this time the rate of release of the herbicide failed to make up for losses in the soil. This time can be related to the initial loading of the herbicide on the polymer substrate from a consideration of the rate of degradation of the linkage. For essentially water-insoluble backbones, the duration of activity will be proportional to the logarithm of the amount of polymer. Thus, half-lives can be calculated for any particular pesticide–polymer for a given set of conditions. For each of the range of combinations of 2,4-D given in Table 1, a comparative half-life has been measured. The period of protection, assuming a threshold dosage, can also be calculated. These data for laboratory conditions have been determined for an extensive range of polymers.[87]

4.3. Release Rates Determined under Field Conditions

The field trial yield is the ultimate test of any agricultural treatment, including that based on controlled delivery. Estimates of the efficiency of any controlled delivery system under field conditions are difficult without quantitative assessments of release. However, this has been rarely reported and data usually repre-

TABLE 1. The Release Characteristics of Some Polymeric Derivatives of 2,4-D.

Method of Preparation	Linkage	Saponification Equivalent[a] (g polymer)	Maximum Releasable 2,4-D (% w/w)	Approx. Half-Life in Soil (days)	Persistence of 2,4-D 10 mg/pot (days)
A Homopolymerization of 2,4–D vinyl ester	–COO–	246[b]	89	–	–
B Copolymerization of 2,4–D vinyl ester and acrylic acid	–COO–	–	63[c]	10	32
C Esterified bark[d]	–COO–	–	12	24	83
D Esterified cellulose[d]	–COO–	–	7	13	50
E Esterified PVA[d,e]	–COO–	1956	45	19	64
F Acylation[f] of low molecular weight PEI[g]	–CONH–	–	87	18	60
G Acylation[h] of medium molecular weight PEI	–CONH–	–	85	15	50
H Acylation[h] of medium molecular weight PEI	–CONH–	–	68	7	25
J 2,4–D acid	–	–	100	11	47
L 2,4–D acid (exposed to rainfall)	–	–	100	7	22

[a] Determined by hydrolysis with alcoholic potassium hydroxide.
[b] Intrinsic viscosity 0.21.
[c] Equivalent to a molar ratio of three carboxylic acid to two 2,4–D ester pendent groups.
[d] Using 2,4–D acid chloride.
[e] PVA = polyvinyl alcohol.
[f] Using 2,4–D acid.
[g] PEI = polyethylenimine.
[h] Using 2,4–D methyl ester.

sent efficacy only, in terms of control of the pest and harvest yields. Methods employed depend very much on the biological response of the pest organism controlled and the pest control situation. For example, the evaluation of nitrogen release from ureaform and isobutylidene diurea (IBDU) was carried out in turf grass by weekly clipping.[112] Microencapsulated chlorpropham was field bioassayed once or twice a week using a given volume of *Lolium multiflorum* Lam. seed.

Control of "fruit rot" with lignin-PCNB combination was assessed in the field periodically by leaving fresh cucumbers on the soil for 5 days covered with a polyethylene sheet. The cucumbers were then removed and examined for fruit rot lesions.[33] Controlled-release formulations of diazinon and acephate used for a range of insect pests on vegetables were not periodically assessed but were evaluated only on the basis of harvest damage and yields.[113]

The assessment of release rates under field conditions is made complex by the often long delivery path between the releasing formulation and the site of pest attack. This is particularly true for systemic pesticides designed to protect plants from foliar attack. In this case the important variable is the concentration of the pesticide in the foliage and not in the immediate area (e.g., the root zone) of the formulation. In addition, the pesticide must be available to the attacking pest.

Rice plants treated with controlled-release systemic insecticides have been bioassayed for periods following treatment by placing small cages around randomly selected rice plants in the field. A number of laboratory-reared adult brown planthoppers were then inserted into the cage, and percent mortality was assessed after 48 h.[114] As the response of the cultured insects to the test insecticide was known, a reliable field bioassay was feasible.

Where possible, field bioassays are best combined with measurements of loss of active ingredient from the formulation. A situation where this has been carried out is in protection of *Cedrela odorata* (Spanish cedar), using systemic insecticides formulated in fatty polyamide pellets, against continual attacks of the mahogany shootborer *Hypsipyla grandella* Zeller.[10] Shootborer attack was assessed every 2 weeks, and the frequency of attack for two different field sites combined is shown in Fig. 11, comparing the performance of five systemic insecticides. Samples of the insecticide–polymer pellets were retrieved at intervals and analyzed for residual insecticide. The results are given in Table 2. At the termination of its insecticidal activity (1 year) the most effective combination (carbofuran) still retained 70% of the original insecticide. By interpolation of the data in Table 2 it can be shown that the insecticide contents of the other controlled-release formulations (except Isolan) were at similar levels at the expiration of effective control. Thus, those formulations with higher rates of release provided shorter periods of protection, and long-term efficacy depended as much on the formulation design as on the inherent insecticidal activity of the active ingredient.

REFERENCES

1. R. M. Wilkins, in R. Baker, Ed., *Controlled Release of Bioactive Materials*, Academic, New York, 1980, p. 343.
2. R. M. Wilkins, in J. S. McLaren, Ed., *Chemical Manipulation of Crop Growth and Development*, Butterworths, London, 1982, p. 111.
3. G. G. Allan, J. W. Beer, M. J. Cousin, and R. A. Mikels, in A. F. Kydonieus, Ed., *Con-*

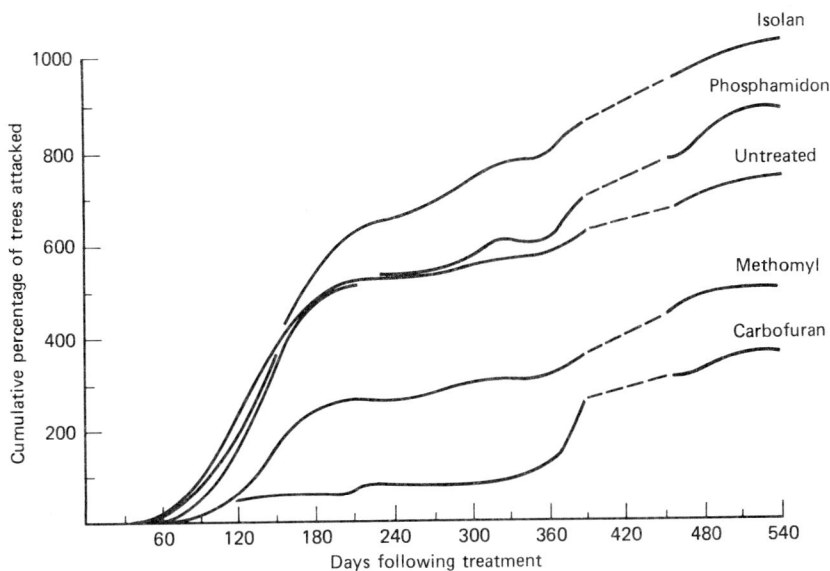

FIGURE 11. Frequency of seedling trees (*Cedrela odorata*) attacked by *H. grandella* larvae following root-zone treatment with controlled-release insecticide pellets. The frequency is expressed in terms of the accumulated percent of trees in each treatment showing any new attack as observed every two weeks.

TABLE 2. The Insecticide Contents of Some Controlled Release Polyamide Matrix Combinations Retrieved From Tropical Soils at Differing Periods After Placement.

Insecticide	Purity of the Technical Insecticide (%)	Insecticide Content of Combination (%)			
		Before Insertion in Soil	After 7 Months[a] in Soil	After 12months[b] in Soil	After 18 Months[c] in Soil
Carbofuran	97.1	24.9	19.0±2.3	17.2±2.3	16.5±2.7
Methomyl	87.8[d]	25.0	14.8±1.0	7.0±3.0	7.4±3.0
Monocrotophos	96.7[d]	25.0	—	10.2±1.5	7.3±1.5
Phosphamidon	90.4[d]	26.8	—	10.0±1.8	6.3±2.0
Isolan	-[e]	25.0	6.3	—	—

[a] Average of two samples.

[b] Average of four samples.

[c] Average of 10 samples.

[d] Determined by elemental analysis (for S,P, or Cl).

[e] Not determined.

173

trolled Release Technologies: Methods, Theory and Applications, Vol. II, CRC Press, Boca Raton, FL, 1980, p.7.

4. N. F. Cardarelli, in N. F. Cardarelli, Ed., *Proc. Int. Controlled Release Pesticide Symp., Wright State Univ., Dayton, OH*, 1975, p. 349.

5. R. W. Baker and H. K. Lonsdale, in N. F. Cardarelli, Ed., *Proc. Int. Controlled Release Pesticide Symp., Wright State Univ., Dayton, OH*, 1975, p. 9.

6. T. J. Roseman, in R. L. Goulding, Ed., *Proc. Int. Controlled Release Pesticide Symp., Oregon State Univ., Corvallis*, 1977, p. 403.

7. S. K. Chandrasekaran, R. W. Baker, R. G. Buckles, and A. S. Michaels, in N. F. Cardarelli, Ed., *Proc. Int. Controlled Release Pesticide Symp., Univ. Akron, OH*, 1974, p. 11.1.

8. G. S. Hartley and I. J. Graham-Bryce, *Physical Principles of Pesticides Behaviour*, Vols. I and II, Academic, London, 1980.

9. G. G. Allan, J. W. Beer, and M. J. Cousin, in H. B. Scher, Ed., *Controlled Release Pesticides*, Am. Chem. Soc., Washington, DC, 1977, p. 94.

10. G. G. Allan, R. I. Gara, and R. M. Wilkins, *Int. Pest. Control*, 16, 4 (1974).

11. G. G. Allan, C. S. Chopra, J. F. Friedhoff, R. I. Gara, M. W. Maggi, A. N. Neogi, S. C. Roberts, and R. M. Wilkins, *Chem. Technol.*, 3, 171 (1973).

12. G. G. Allan, J. F. Friedhoff, W. J. McConnel, and J. C. Powell, *J. Macromol. Sci. Chem.*, A10 (1, 2), 223 (1976).

13. H. M. Flint, L. M. McDonough, S. S. Salter, and S. Walters, *J. Econ. Entomol.*, 72, 798 (1979).

14. L. I. Butler and L. M. McDonough, *J. Chem. Ecol.*, 5, 825 (1979).

15. B. A. Bierl-Leonhardt, E. D. DeVilbiss, and J. R. Plimmer, *J. Econ. Entomol.*, 72, 319 (1979).

16. M. Beroza, B. A. Bierl, P. James, and E. D. DeVilbiss, *J. Econ. Entomol.*, 68(3), 369 (1975).

17. J. J. Karr, *J. Assoc. Off. Anal. Chem.*, 63(5), 999 (1980).

18. R. J. Argauer, *J. Assoc. Off. Anal. Chem.*, 63(5), 1003 (1980).

19. T. D. Fitzgerald, A. D. St. Clair, G. E. Daterman, and R. G. Smith, *Environ. Entomol.*, 2(4), 607 (1973).

20. R. A. Stokes, J. R. Coppedge, D. L. Bull, and R. L. Ridgway, *J. Agric. Food Chem.*, 21(1), 103 (1973).

21. F. W. Harris, in A. F. Kydonieus, Ed., *Controlled Release Technologies: Methods, Theory and Applications*, Vol. II, CRC Press, Boca Raton, FL, 1980, p. 63.

22. R. M. Wilkins, in *Proc. BCPC Symp. Persistence Insecticides Herbicides*, Monograph No. 17, Croydon, 1976, p. 201.

23. G. G. Allan, C. S. Chopra, A. N. Neogi, and R. M. Wilkins, *TAPPI*, 54(8), 1293 (1971).

24. G. G. Allan, J. W. Beer, M. J. Cousin, W. J. McConnell, J. C. Powell, and A. Yahiaoui, in L. G. Donaruma and O. Vogl, Eds., *Polymeric Drugs*, Academic, New York, 1978, p. 193.

25. G. G. Allan, C. S. Chopra, A. N. Neogi, and R. M. Wilkins, *Nature*, 234, 349 (1971).

26. R. M. Wilkins, in N. F. Cardarelli, Ed., *Proc. Int. Controlled Release Pesticide Symp., Univ. Akron, OH*, 1976, p. 7.1.

27. C. L. Mehltretter, W. B. Roth, F. B. Weakley, T. A. McGuire, and C. R. Russell, *Weed Sci.*, 22(5), 415 (1974).

28. R. de B. Ashworth, J. Henriet, J. F. Lovett, and A. Martijn (Compilers), *Analysis of Technical and Formulated Pesticides*, CIPAC Handbook Vol. I and IA, Collaborative International Pesticides Analytical Council, Harpenden, V.K., 1980.

29a. G. Zweig, Ed., *Analytical Methods for Pesticides and Plant Growth Regulators*, Vols. I-XI, Academic, New York, 1963-1980.

  b. K. G. Das, Ed., *Pesticide Analysis*, Marcel Dekker, New York, 1981.

30. H. B. Scher, in H. B. Scher, Ed., *Controlled Release Pesticides*, Am. Chem. Soc., Washington, DC, 1977, p. 126.

31. M. M. Schreiber and M. D. White, *Weed Sci.* **28**(6), 685 (1980).

32. R. M. Wilkins, *Pesticide Sci.*, **9**, 443 (1978).

33. H. T. Dellicolli, in A. F. Kydonieus, Ed., *Controlled Release Technologies; Methods*, Washington, DC, 1977, p. 84.

34. H. T. Dellicolli, in A. F. Kydonieus, Ed., *Controlled Release Technologies; Methods, Theory and Applications*, Vol. II, CRC Press, Boca Raton, FL, 1980, p. 225.

35. G. G. Allan and A. N. Neogi, U.K. Patent 1502441 (1978).

36. H. E. Thompson, G. G. Allan, and A. N. Neogi, *Int. Pest. Control*, **23**, 10 (1981).

37. R. M. Wilkins, in G. H. Williams, Ed., *Proc. Crop Protection in Northern Britain*, British Crop Protection Council, Croydon, 1981, p. 261.

38. R. M. Wilkins, unpublished data, 1980.

39. E. R. Mitchell, Ed., *Management of Insect Pests with Semiochemicals*, Plenum, New York, 1981.

40. D. A. Nordlund and R. L. Jones, *Semiochemicals, Their Role in Pest Control*, Wiley, New York, 1981.

41. D. R. Lauren, *Environ. Entomol.*, **8**, 914 (1979).

42. C. Sheppard, *J. Econ. Entomol.*, **73**, 276 (1980).

43. J. Weatherston, M. A. Golub, T. W. Brooks, Y. Y. Huang, and M. H. Benn, in E. R. Mitchell, Ed., *Management of Insect Pests with Semiochemicals*, Plenum, New York, 1981, p. 425.

44. J. H. Caro, B. A. Bierl, H. P. Freeman, D. E. Glotfelty, and B. C. Turner, *Environ. Entomol.*, **6**(6), 877 (1977).

45. J. H. Cross, *J. Chem. Ecol.*, **6**(4), 781 (1980).

46. B. A. Bierl-Leonhardt, E. D. DeVilbiss, and J. R. Plimmer, *J. Econ. Entomol.*, **72**, 319 (1979).

47. D. L. Bull, J. R. Coppedge, R. L. Ridgway, D. D. Hardee, and T. M. Graves, *J. Econ. Entomol.*, **2**(5), 829 (1973).

48. E. Ashare, T. W. Brooks, and D. W. Swenson, in D. R. Paul and F. W. Harris, Ed., *Controlled Release Polymeric Formulations*, Am. Chem. Soc., Washington, DC, 1976, p. 273.

49. T. C. Baker, R. T. Cardé, and J. R. Miller, *J. Chem. Ecol.*, **6**, 749 (1980).

50. B. A. Bierl and D. DeVilbiss, in F. W. Harris, Ed., *Proc. Int. Controlled Release Pesticide Symp.*, *Wright State Univ., Dayton, OH*, 1975, p. 230.

51. A. F. Kydonieus, in H. B. Scher, Ed., *Controlled Release Pesticides*, Am. Chem. Soc., Washington, DC, 1977, p. 152.

52. J. H. Cross, J. H. Tumlinson, R. E. Heath, and D. E. Burnett, *J. Chem. Ecol.*, **6**(4), 759 (1980).

53. J. H. Caro, D. E. Glotfelty, and H. P. Freeman, *J. Chem. Ecol.*, **6**(1), 229 (1980).

54. D. G. Campion et al., in E. R. Mitchell, Ed., *Management of Insect Pests with Semiochemicals*, Plenum, New York, 1981.

55. J. H. Cross, *J. Chem. Ecol.*, 6(4), 789 (1980).

56. J. H. Cross, E. R. Mitchell, J. H. Tumlinson, and D. E. Burnett, *J. Chem. Ecol.*, 6(4), 771 (1980).

57. H. Arn, in F. J. Ritter, Ed., *Chemical Ecology: Odour Communications in Animals*, Elsevier/North Holland Biomedical Press, Amsterdam, 1979, p. 365.

58. H. Tashiro, M. J. Gieselmann, and W. L. Roelofs, *Environ. Entomol.*, 8, 931 (1979).

59. F. J. Ritter, Ed., *Chemical Ecology: Odour Communication in Animals*, Elsevier/ North Holland Biomedical Press, Amsterdam, 1979.

60. D. L. Overhulser, G. E. Daterman, L. L. Sower, C. Sartwell, and T. W. Koerber, *Can. Entomol.*, 112, 163 (1980).

61. J. R. Plimmer, B. A. Bierl, R. E. Webb, and C. P. Schalbe, in H. B. Scher, Ed., *Controlled Release Pesticides*, Am. Chem. Soc., Washington, DC, 1977, p. 168.

62. J. O. Schmidt and W. D. Seabrook, *J. Econ. Entomol.*, 72, 509 (1979).

63. A. B. Broce, J. L. Goodenough, and J. W. Snow, *Environ. Entomol.*, 8, 824 (1979).

64. T. W. Brooks, C. C. Doane, D. G. Osborn, and J. K. Haworth, in R. Baker, Ed., *Controlled Release of Bioactive Materials*, Academic, New York, 1980, p. 227.

65, J. R. Plimmer and M. N. Inscoe, in F. J. Ritter, Ed., *Chemical Ecology: Odour Communication in Animals*, Elesevier/North Holland, Amsterdam, 1979, p. 249.

66. L. Burgess and J. E. Wiens, *Can. Entomol.*, 112, 93 (1980).

67. H. Den Ouden and J. Theunissen, *Neth. J. Plant Pathol.*, 86, 17 (1980).

68. Idemitsu Kosan Co. Jap. Patent 8081802 (20 June 1980).

69, P. F. Prevett and S. M. Blatchford, *Trop. Stored Prod. Inf.*, 23, 6 (1972).

70. D. P. Childs and J. E. Overby, *J. Stored Prod. Res.*, 15, 123 (1979).

71. R. B. Bruce, A. J. Robbins, and T. O. Tuft, *J. Agric. Food Chem.*, 10, 18 (1962).

72. C. H. Bell, *J. Stored Prod. Res.*, 15, 53 (1979).

73. A. Tremblay and M. Murgett, *J. Econ. Entomol.*, 72, 616 (1979).

74. N. F. Cardarelli, in A. F. Kydonieus, Ed., *Controlled Release Technologies: Methods, Theory and Applications*, Vol. II, CRC Press, Boca Raton, FL, 1980, p. 55.

75. A. R. Quisumbing, A. F. Kydonieus, D. R. Calsett, and J. B. Haus, in N. F. Cardarelli, Ed., *Proc. Int. Controlled Release Pesticide Symp., Univ. Akron, OH*, 1976, p. 3.40.

76. A. F. Kydonieus, Ed., *Controlled Release Technologies: Methods, Theory and Application, Vol. I, CRC Press, Boca Raton, FL, 1980, p. 235.*

77. N. Cardarelli, *Controlled Release Pesticides Formulations*, CRC Press, Boca Raton, FL, 1976, p. 101.

78. A. F. Kydonieus, Ed., *Controlled Release Technologies: Methods, Theory and Applications*, Vol. I, CRC Press, Boca Raton, FL, 1980, p. 15.

79. S. Z. Mansdorf, in N. Cardarelli, Ed., *Proc. Int. Controlled Release Pesticides Symp., Univ. Akron, OH*, 1974, p. 12.1.

80. G. A. Janes and S. Z. Mansdorf, in R. L. Goulding, Ed., *Proc. Int. Controlled Release Pesticides Symp., Oregon State Univ. Corvallis*, 1977, p. 11.

81, G. A. Janes, in F. W. Harris, Ed., *Proc. Int. Controlled Release Pesticide Symp., Wright State Univ., Dayton, OH*, 1975, p. 326.

82. G. A. Janes, in *Proc. Int. Symp. Controlled Release of Bioactive Materials*, Nat. Bur. Standards, Gaithersburg, MD, 1978, p. 3.1.

83. F. W. Harris, M. R. Dykes, and A. E. Aulabaugh, in R. L. Goulding, Ed., *Proc. Int. Controlled Release Pesticide Symp., Oregon State Univ., Corvallis*, 1977, p. 1.

84. M. B. Shambhu, G. A. Digenis, D. K. Gulati, K. Bowman, and P. S. Sabharwal, *J. Agric. Food Chem.*, **24***(3), 666 (1976)*.

85. G. G. Allan and A. N. Neogi, *Int. Pest Control*, **14**, 21(1972).

86. S. Bittner and I. Perry, *Chim. Ind. (Milan)*, **61**(4), 291 (1979).

87. A. N. Neogi and G. G. Allan, in A. C. Tanquary and R. E. Lacey, Eds., *Controlled Release of Biologically Active Agents*, Plenum, New York, 1974, p. 195.

88. R. M. Wilkins and J. Williams, unpublished data, 1977.

89. J. R. Coppedge, R. A. Stokes, R. L. Ridgway, and D. L. Bull, *J. Econ. Entomol.*, **68**(4), 508 (1975).

90. J. R. Baur, *J. Environ Qual.*, **9**(3), 379 (1980).

91. A. E. Smith and B. P. Verma, *Weed Sci.* **25**(2), 175 (1977).

92. C. R. Worthing, Ed., *The Pesticide Manual*, 6th ed., British Crop Protection Council, Croydon, 1979.

93. R. M. Wilkins, Ph.D. thesis, Univ. Washington, Seattle, WA, 1972.

94. F. W. Knapp and C. Nontapan, in R. Baker, Ed., *Controlled Release of Bioactive Materials*, Academic, London, 1980, p. 267.

95. R. M. Wilkins and S. Batterby, unpublished data, 1980.

96. R. J. Hance, Ed., *Interactions between Herbicides and the Soil*, Academic, London, 1980.

97. J. Feldmesser and B. S. Shasha, in R. L. Goulding, Ed., *Proc. Int. Controlled Release Pesticide Symp., Oregon State Univ., Corvallis*, 1977, p. 205.

98. J. Feldmesser, B. S. Shasha, and W. M. Doane, in N. F. Cardarelli, Ed., *Proc. Int. Controlled Release Pesticide Symp., Univ. Akron, OH*, 1976, p. 6.18.

99. M. M. Schreiber, B. S. Shasha, M. A. Ross, P. L. Orwick, and D. W. Edgcomb, Jr., *Weed Sci.*, **26**(6), 679 (1978).

100. J. A. Bakan, in F. W. Harris, Ed., *Proc. Int. Controlled Release Pesticide Symp., Wright State Univ., Dayton, OH*, 1975, p. 76.

101. T. Dappert and C. Thies, in *Proc. Int. Symp. Controlled Release of Bioactive Materials*, Nat. Bur. of Standards, Gaithersburg, MD, 1978, p. 2.18.

102. J. Fiskill and R. B. Diamond, in R. L. Goulding, Ed., *Proc. Int. Controlled Release Pesticide Symp., Oregon State Univ., Corvallis*, 1977, p. 299.

103. E. J. Holcomb. *Commun. Soil Sci. Plant Anal.*, **12**(3), 271 (1981).

104. F. W. Harris, A. E. Aulabaugh, R. D. Case, M. K. Dykes, and W. A. Feld, in D. R. Paul and F. W. Harris, Eds., *Controlled Release Polymeric Formulations*, Am. Chem. Soc., Washington, DC, 1976, p. 222.

105. F. W. Harris, M. R. Dykes, J. A. Baker, and A. E. Aulabaugh, in H. B. Scher, Ed., *Controlled Release Pesticides*, Am. Chem. Soc., Washington, DC, 1977, p. 102.

106. G. G. Allan, J. W. Beer, and N. J. Cousin, in H. B. Scher, *Controlled Release Pesticides*, Am. Chem. Soc., Washington, DC, 1977, p. 94.

107. S. Bittner, I. Perry, and Y. Knobler, *Phytochem.*, **16**, 305 (1977).

108. B. Shasha, in *Proc. Int. Symp. Controlled Release Bioactive Materials*, Nat. Bur. Standards, Gaithersburg, MD, 1978, p. 2.31.

109. C. L. McCormick and M. Fooladi, in H. B. Scher, Ed., *Controlled Release Pesticides*, Am. Chem. Soc., Washington, DC, 1977, p. 112.

110. K. Savage, C. McCormick, and B. Hutchinson, *Proc. Int. Symp. Controlled Release of Bioactive Materials*, Nat. Bur. Standards, Gaithersburg, MD, 1978, p. 318.

111. K. G. Das, in *Proc. Int. Symp. Controlled Release of Bioactive Materials*, Nat. Bur. Standards, Gaithersburg, MD, 1978, p. 3.39.

112. F. Waddington, in R. L. Goulding, Ed., *Proc. Int. Controlled Release Pesticide Symp.*, Oregon State Univ., Corvallis, 1977, p. 319.

113. N. Gauthier, in *Proc. Int. Symp. Controlled Release of Bioactive Materials*, Nat. Bur. Standards, Gaithersburg, MD, 1978, p. 5.29.

114. R. M. Wilkins and E. A. Heinrichs, in preparation.

CHAPTER SEVEN

Environmental Aspects of Controlled-Release Pesticides

G. ZWEIG
School of Public Health
University of California
Berkeley, California

CONTENTS

1. INTRODUCTION

In this chapter, we shall discuss the possible impact of environmental factors on controlled-release pesticide formulations. We shall examine the physicochemical and biological fate and movement of these products and, conversely, the possible adverse effect of controlled-release formulations on nontarget organisms in the environment. Widespread environmental effects uniquely due to controlled-release pesticide formulations different from those of conventional formulations have not been observed in the "real world," probably for two main reasons: (1) The matrix of controlled-release products is most often biologically and chemically inert and, therefore, not degraded to possible toxic products. (2) controlled-release products are not yet as widely used as conventional pesticide formulations and, therefore, outside of the laboratory no massive impacts have been either observed or reported.

That some controlled-release products are not completely inert and do have adverse environmental effects on selected nontarget organisms is discussed in Section 3 of this chapter. The governmental regulatory agency responsible for the registration of pesticides in the United States, the Environmental Protection Agency (EPA), is aware of these potential problems and is wrestling with the task of establishing guidelines for controlled-release pesticide formulations. Prior laboratory and limited field tests will prevent the introduction of chemicals into the environment that pose unreasonably adverse hazards to nontarget organisms. The philosophy underlying the establishment of these registration guidelines is discussed in Section 4.

The three major advantages of controlled-release pesticide formulations are

1. The sparing effect of the controlled-release phenomenon and thus the

more economical use of toxicants to achieve maximum control of the target pest represents an economic and toxicological advantage.

2. The environmental instability of several pesticide chemicals, for example, insect growth regulators and pheromones, can be overcome by incorporating them into a controlled-release matrix, thus stabilizing the chemical and releasing it slowly over an extended period of time for prolonged biological activity.

3. Aquatic herbicides and molluscicides are most effective when released slowly and over a long period of time. If these pesticides were not incorporated into a controlled-release formulation, they could not be used on a practical scale for pest control.

These topics have been expanded in other chapters throughout this book. It is the purpose of this chapter to examine the possible adverse environmental impacts due to controlled-release pesticide formulations and to suggest possible remedial action where these problems occur. Governmental requirements for environmental tests in the laboratory and field will enhance the predictive capability of regulators within the EPA and suggest corrective measures to the manufacturer before these new controlled-release pesticide products are widely disseminated in the landscape.

2. ENVIRONMENTAL FATE OF CONTROLLED-RELEASE PESTICIDE FORMULATIONS

This chapter will focus on the unique combination of the biocide–controlled-release matrix and not on the biocide itself. In almost all cases where a controlled-release pesticide product has been developed, a host of environmental and toxicological tests will have been performed already in order to obtain registration in the country of use. However, particular questions for controlled-release pesticide formulations that may not be obvious offhand will have to be answered, as for example: Does the active ingredient (pesticide or biocide) behave differently when formulated in a controlled-release matrix (plastic or polymer)? These questions have been recently studied and reviewed very carefully by Cardarelli and Walker,[1] who concluded that a great deal is known about the effect of environmental factors on the organic polymers making up the controlled-release product. It is difficult to visualize significant deleterious effects of controlled-release pesticide formulations on nontarget biota in the environment. Nevertheless, some knowledge must be garnered on the rate of degradation of the polymers and the identification of breakdown products. An attempt should be made to assess the possible environmental impact of the polymer itself and its breakdown products on nontarget organisms.

Because Cardarelli and Walker[1] were able to place the potential environmental hazard of polymers used in controlled-release formulations into the correct perspective, a paragraph from their publication is quoted here verbatim:

It should be recognized that the polymeric material must degrade in some fashion before there can be any environmental impact in the chemical, biochemical or biological sense. With reference to soil or the beds of watercourses, it is conceivable, though barely, that the addition of a polymer to the ambient environment will result in physical changes. It is well known that the application of an inert material to soil aids in aeration. Since the authors cannot visualize more than a few hundred pounds per acre of a given polymer being applied annually (a few pounds per acre being much more likely), and since the weight of an acre-foot of soil is in excess of a million pounds, the environmental impact after dilution would appear to be nil. If polymers for use in controlled release were completely inert—or their degradation rate was measured in geologic time—as occurs with glass, the cumulative aspect would be a matter of concern. However, all organic polymers degrade upon exposure of the outdoor environment, and with few exceptions their nature is completely obliterated after a number of decades.

That this author does not fully subscribe to Caradarelli and Walker's unconcern will become apparent in subsequent sections of this chapter.

If polymers do indeed degrade environmentally, the following natural factors may be independently or collectively responsible for the breakdown: solar radiation (ultraviolet), heat, hydrolysis, oxidation, and biological action. In conventional polymer technology it is usually the practice to add stabilizers to enhance the environmental stability of the polymer. However, in controlled-release pesticide design it might be more prudent to delete the stabilizer, or, indeed, to add materials that enhance the rate of environmental degradation. A description of these additives is dealt with in other parts of this book.

To examine the environmental breakdown of controlled-release polymers, a brief section on each of the most important materials will illustrate potential environmental problems. Where no such studies have been performed, sometimes an evaluation can be attempted on the basis of analogous structure. The information given here is based in large part on an EPA-sponsored project and its final report.[1]

2.1. Natural and Synthetic Rubbers

Synthetic rubber is basically polyisoprene with some impurities like unreacted monomer, catalytic fragments, and so on, whereas natural rubber, also polyisoprene, contains many other impurities, such as fatty acids, sterols, and esters. Natural rubber degrades rapidly due to physicochemical reactions and bacterial and fungal attack. Addition of sodium sulfite, bactericides, and fungicides retards the biological degradation of natural rubber.

Natural antioxidants are destroyed during vulcanization of rubber, leading to oxidation products including aldehydes, ketones, alcohols, and ethers. Ozone, in the presence of ultraviolet radiation, reacts with double bonds, resulting in the formation of ozonides and subsequent chain scission. Natural rubber is ultimately degraded to nontoxic, natural constituents. However, there might be some concern about the possible toxicologically significant intermediates of decomposing vulcanized rubber, including sulfonated aromatic compounds, mercaptans, alcohols, ketones, and aldehydes. None of these intermediates, however, appears to have been systematically identified.

Synthetic rubber is less prone to microbial attack than natural rubber, but its occurrence at a slow rate has been reported.[2] On the other hand, rubber formulations can be modified for faster deterioration by increasing the fatty acid content of the final product. The rate of environmental degradation of rubber and the nature of the degradation products depend on the type of additives and the characteristics of the polymers. It is necessary, therefore, to study the environmental characteristics of a given system and not just the polymer alone. Chlorinated natural rubber, which is used in antifouling coatings, is highly resistant to microbial attack, but it may be toxic to a number of microorganisms (N. F. Cardarelli, unpublished observations).

2.2. Polyisobutylene and Butyl Rubbers

Compounded butyl rubber materials are resistant to attack by moisture and oxidation. Since butyl rubber is a copolymer of isobutylene and isoprene, the ratio of the two monomers will have an effect on the environmental resistance of the final product. The degree of environmental persistence may also depend on the type of curing agent used, for example, sulfur or lead peroxide. Butyl rubber is attacked and ingested by termites but is fairly resistant to bacterial attack.[3] In summary, butyl and chlorobutyl elastomers have a high degree of resistance to oxidation and microbial attack and therefore are not expected to adversely affect the environment.

2.3. Ethylene-Propylene Polymers

These synthetic polymers are much more resistant to microbial and physicochemical attack. The manufacturer of these elastomers (B. F. Goodrich) was unable to supply information on their degradation. Qualitatively, however, ethylene-propylene polymers show excellent resistance to attack by ozone, oxygen, heat, and conditions leading to hydrolysis. If these polymers are used for controlled-release formulations, no significant amount of environmental degradation is expected, and adverse effects on the environment are believed to be minimal.

2.4. Styrene-Butadiene Copolymers

Generally, these polymers are readily attacked by oxygen, ozone, and ultraviolet radiation, and protectants, such as 2,6-di-*tert*-butyl-4-methylphenol, are added during the manufacturing process. Due to cross-linkage of butadiene, upon exposure to ionizing radiation and/or oxygen, the copolymeric material becomes brittle. When these copolymers are emulsified during processing with a fatty acid, they become more resistant to bacterial degradation. No information was found in the literature on the identification of breakdown products due to microbial activity.

2.5. Polyacrylonitriles

These polymers tend to be fairly resistant to microbial attack and thermal degradation. However, ozone rapidly degrades them.

2.6. Polyacrylates

Carboxylated acrylic systems such as Carboset are rapidly degraded by microorganisms, especially fungi. Carboset polymers are ingested by mulluscs, insects, and probably nematodes (N. F. Cardarelli, personal observation). No information was found in the literature on the toxicity of Carboset to fish or plants or other segments of the biota; also, no environmental studies have been published.

2.7. Polyesters

Polyesters are noted for their resistance to chemical and radiation-induced degradation. Hydrolysis can occur at elevated temperatures, leading to some degree of depolymerization. Degradation due to solar radiation depends on the type of polyester. For example, repeated methylene chain units produce the most stable polyester, whereas the introduction of carboxyl groups leads to decreased stability. Breakdown products at elevated temperatures include allyl alcohol, formaldehyde, acetaldehyde, carbon dioxide, carbonate ions, and formic and acetic acids. No information was found on the biodegradation of these polymers.

2.8. Polyamides

This group of polymers is widely used in controlled-release pesticide formulations, for example, Penncap-M and Altosid-SR-10. Therefore, the environmental fate of this capsule material is of great interest. Polyamides degrade under the influence of heat and radiation to yield simple compounds like hydrocarbons, cyclopentanone, CO_2, CO, and water. Under ambient conditions, however, the degradation of nylon microcapsules is believed to be very slow. Although no environmental studies have been reported, the EPA, nevertheless, granted exemption from the requirement to establish a food tolerance for capsules made of cross-linked nylon-type polymer[4] containing methyl parathion (Penncap-M) and for the polyamide polymer derived from sebacic acid, vegetable oil acids with or without dimerization, terephthalic acid, and/or ethylenediamine for methoprene only.[5] The reason for the latter restriction (methoprene only) was that polyamide polymers may extend the life of the active pesticidal ingredient, as they are indeed intended to do, and this may cause illegal (overtolerance) residues of the pesticide in the food crop.

2.9. Polyethylene

Polyethylenes, including chlorinated and chlorosulfonated polyethylenes, are even more resistant to environmental degradation than chloroprene rubber. These materials are also highly resistant to microbial attack and are generally chemically inert.

2.10. Polyvinyl Chloride

The microbiological degradation of PVC depends on the type of plasticizer used. Phthalate and phosphate plasticizers increase resistance to biological attack, whereas plasticizers like adipate, azelate, and sebacate tend to increase microbial degradation. Most PVCs degrade from exposure to ultraviolet radiation. Thermal degradation leads to dehydrochlorination.

2.11. Chloroprene Polymers

The presence of chlorine in neoprene renders the material very resistant to microbial attack.[6] Environmental degradation due to sunlight and oxidation does not occur to any appreciable extent.

2.12. Polyurethanes

Amide and amino groups impart a high degree of biodegradability to polyurethanes.[6] Polyester-based polyurethanes are equally prone to hydrolysis, while polyether-based elastomers are highly resistant to hydrolysis. Polyurethanes are resistant to chemical attack.

2.13. Acetal Copolymers

Acetal copolymers, such as Celcon, a formaldehyde-ethylene oxide material, are used in controlled-release dispensers, such as Conrel. These copolymers degrade in soil or water to formaldehyde and ethylene oxide. They also degrade rapidly under ultraviolet irradiation, and hydrolysis occurs slowly in the presence of moisture.

2.14. Cellulosic Materials

Cellulosic materials in general degrade rapidly when exposed to natural forces. For example, cellophane is degraded rapidly by sunlight and microorganisms, especially in loose, moist soil. Chemical substitution of hydrogen by alkoxy or hydroxylalkoxyl groups enhances the resistance to microbial attack. Cellulosic materials and their breakdown products are generally considered essentially nontoxic to humans, and therefore, are not expected to pose an environmental hazard when used as matrices for controlled-release pesticide formulations.

2.15. Other Materials

Although silicone rubber is not presently used in controlled-release pesticide formulations, it has found use in controlled-release drug-delivery systems and thus may find future application in pesticide products. Silicones are recognized as nontoxic and extremely inert to biological attack.

Polyacrylamides are not presently used for controlled-release formulations. These materials degrade slowly under the influence of heat, and it is anticipated that they will gradually depolymerize to ammonia or ammonium ions depending on pH. Hydrin rubber represents another controlled-release matrix for pesticides; it is the homopolymer of epichlorohydrin. Hydrin rubber is stable to heat and ozone but might be susceptible to microbial attack.

Microporous Plastic Sheet® is made of silica-modified PVC or vinyl chloride-vinyl acetate copolymer. It is a likely candidate for controlled-release formulations, but little is known about its environmental behavior.

Polyoxymethylene copolymer, when used as an inert controlled-release dispenser for the insect attractant gossyplure, has been exempted by the EPA from the requirement to establish a residue tolerance on cottonseed.[7] Cross-linked polyurea-type encapsulating material for pesticide formulations prior to planting has also been exempted from these requirements.[8] By granting these exemptions, the EPA must believe that these polymeric materials do not pose an unreasonable hazard to the environment, including edibles, when used under the specified conditions.

2.16. Polymer Additives

Space does not permit extensive coverage of the nature and environmental impact of polymer additives. These compounds include a variety of plasticizers, antioxidants, and release regulators, and many of them are exempted from the requirement to establish a tolerance in raw agricultural crops (see Section 2.15). Although some additives like the phthalates have been considered to present an environmental problem, one would have to assess the escape of phthalates from controlled-release products to evaluate possible environmental impact and human exposure.

3. ENVIRONMENTAL IMPACT ON NONTARGET ORGANISMS

A serious concern about controlled-release pesticide formulations is the possibility that they might act on nontarget organisms in an unexpected manner. Chapter 9 covers the toxicology of controlled-release formulations, including chronic vs. acute toxicity. As a rule, and indeed by design, controlled-release pesticide formulations by their very nature are less acutely toxic to warm-blooded animals than conventionally formulated pesticides (emulsifiable concentrates or wettable powders) containing the same active ingredient. For example, the acute LD_{50} of methyl parathion is much greater in rats and mice than that of the microencapsulated product, Penncap-M, as shown in the following table

Acute Oral LD_{50} [9]

Material	Rats	Mice
Methyl parathion	9–25 mg/kg	10–30 mg/kg
Penncap-M	56	>60

The difference of the dermal LD_{50} between the emulsifiable concentrate and microencapsulated methyl parathion is even more striking, 100 mg/kg compared to 1200 mg/kg. Similar factors are expected for humans, and thus the controlled-release pesticides afford a greater degree of safety to pesticide applicators and farmworkers than conventional pesticide formulations.

There appear, however, to be two recorded cases where a controlled-release pesticide formulation has caused or may cause environmental problems, as will be described below.

3.1. Effect of Controlled-Release TBTO on Tropical Food Fish[10]

A slow-release formulation of the molluscicide TBTO [bis(tri-*n*-butyltin)oxide] in a natural rubber compound matrix (Biomet-SRM or MT-1E) was tested against the tropical food fish *Tilapia mossambica.* The 24-h LC_{50} of TBTO was found to be 0.028 ppm. Sublethal concentrations of TBTO released from MT-1E (0.008 and 0.005 ppm) were tested in the laboratory. Fish exposed to the higher concentration lost weight, possibly due to metabolic inhibition. During the first week's exposure, the fish suffered from an eye condition expressed as a whitish opacity at the outermost corneal layer of the eye. It was also observed that the behavior among male animals increased, leading to death. The authors concluded, therefore, that TBTO released from controlled-release formulations has a sublethal effect on this species of tropical food fish. These results suggest that MT-1E might be an environmental hazard to certain beneficial fish species. It would be advisable to conduct laboratory experiments prior to the broad base use of new controlled-release pesticide formulations.

3.2. Effect of Controlled-Release Insecticides on Honeybees

An unexpected environmental hazard from encapsulated methyl parathion was first reported in 1976[11] and since then has been the subject of intensive investigation. The hazards to honeybees[12] and the economic benefits of Penncap-M[13] have been the subject of two papers that will be reviewed in the following section.

Penncap-M insecticide is a controlled-release formulation of methyl parathion encapsulated in nylon-type plastic microcapsules 30–50 μm in diameter. This size is about that of entomophilous pollen.[12] Bees have branched hairs that fit and harvest pollen grains and cannot distinguish between pollen and microcapsules of Penncap-M. The microcapsules adhere to bee hairs more easily than other types of pesticide dust. Barker and co-workers[12] observed that dyed Penncap-M adhered to the legs of bees kept in small cages. It appears, therefore, that bees foraging on crops sprayed with Penncap-M can be contaminated with microcapsules containing a highly toxic insecticide (methyl parathion), which they then store together with pollen in the hive combs.

In actual field studies, residues of 0.14 ppm methyl parathion were found in pollens collected from the hind legs of bees foraging alfalfa that had been sprayed with 0.3 kg/ha active ingredient (methyl parathion) from Penncap-M 48 h earlier.

Penncap-M is also consumed by bees, as demonstrated by the presence of dyed Penncap-M capsules in the midgut of the majority of live and dead bees collected from a plot previously treated with dyed Penncap-M.[13] On the other hand, honey from contaminated colonies contained very little methyl parathion (0.06 ppm) and no capsules. When worker bees and a queen were confined to combs from colonies that had been killed 19 months earlier, presumably from methyl parathion, the adult bees were poisoned. Other contaminated honeycombs from a colony deliberately exposed to blooming rape treated with Penncap-M still retained 97% lethality to exposed bees 10 months later.

The acute oral toxicity of microencapsulated methyl parathion to honeybees is much less than that of the emulsified concentrate product; it is of the same order of magnitude as observed from experiments with rodents. The real hazard from Penncap-M appears to be the longer-lasting effect of the toxic pesticide being liberated from the microcapsules. It has been shown by Barker[12] that residues of Penncap-M remain highly toxic to honeybees 3–5 days after application, whereas alfalfa sprayed with an emulsifiable concentrate of methyl parathion lost most of the toxicity within 12–14 h.

It is believed that another insecticide, carbaryl, applied as dust or in other conventional formulations, also represents a severe hazard to honeybees[14] by being carried back to the hives in significant amounts. Lowell[14] also believes that a potential hazard might even exist from the newer, synthetic pyrethroid insecticides, although no experimental evidence has yet been reported. There appears to be overwhelming evidence that Penncap-M is highly toxic to bees, but it is possible to minimize these adverse effects in a number of ways. One technique consists of soaking the contaminated combs in water for 24 h; the contaminated pollen is washed from the cells, and the combs are dried. Precautionary steps to prevent unnecessary bee kills are listed on the labels of the pesticide container: "Do not apply to weeds in bloom on which an economically significant number of bees are actively foraging." Five major crops in the United States (corn, small grains, cotton, tobacco, and potatoes) are grown without dependence on bee pollination and, therefore, in these crops the hazard from using Penncap-M is minimal.

Other potential remedies for reducing bee kill due to Penncap-M intoxication include the addition of "stickers" or antistatics to the controlled-release capsules, changing capsule size to reduce confusion with pollen, and adding pheromones to capsules to change the forage patterns of honeybees.[13]

3.3. Other Studies on Environmental Impact

It is not intended to make an exhaustive survey of all studies on the environmental impact of controlled-release pesticides but rather to briefly cite illustrative studies that have been specifically conducted to elucidate the environmental fate of these products.

3.3.1. Environmental Fate of Organotin Molluscicides

Cardarelli and Evans[15] have found that trialkyl and triaryl organotins from controlled-release formulations degrade readily in water under various environmental influences (half-life ranges from 4 to 18 days). Hydrolysis follows first-order reaction kinetics and is postulated to proceed from tributyltin fluoride to tributyltin hydroxide, then to tributyl- and dibutyltin oxide. The ultimate decomposition product is SnO_2, which is virtually nontoxic.

In soil, organotins absorb on mineral particulates, which immobilize the activity of these compounds.

It appears that nontarget organisms like fish, crayfish, and aquatic plants are unaffected at concentrations that are lethal to snails (these compounds are efficaceous as molluscicides). Controlled-release technology is keyed to the continuous exposure of the target species (snails) at ultralow concentrations and the consequent terminal, chronic intoxication. A more detailed discussion of these experiments is found in Chapter 9.

3.3.2. Movement of Synthetic Microspheres in Saturated Soil Columns

Lahav and Tropp[16] have studied synthetic cellulose microspheres (1-10 μm) and latex microspheres (0.21-0.12 μm), which were passed through columns of sandy soils (90% sand, 10% clay). Microspheres were very mobile in both heavy and light soils. The larger-diameter spheres were partially retained but migrated, probably due to sedimentation. Repeated washings of the column slowed the movement of the microspheres, probably due to the blocking of the large pores in the soil. Large microspheres, above 8 μm in diameter, did not appear to move vertically in the soil. These experiments demonstrate that encapsulated pesticides are not necessarily passive microspheres, but do move in soils where they may reach the target (e.g., plant roots or nematodes) or may adversely affect the environment by contaminating ground water. These conclusions, however, are highly speculative, and further laboratory and field research studies are essential to elucidate the potential hazard of microencapsulated products.

3.3.3. Evaluation of Controlled-Release Aquatic Herbicides

Janes and Mansdorf[17] have designed a microenvironmental tank to study the fate of elastomer carriers in a laboratory environment. They used a 10-gal aquarium containing fish, snails, hydrosoil, and three aquatic plant species and tested the controlled-release herbicide 2,4-D butoxyethanol ester. These researchers studied the degradation of the rubber matrix and the herbicide, competing rates of release, and the deterioration of the matrix and toxicant.

4. REGULATORY REQUIREMENTS

All pesticides used and distributed in the United States are governed by the Federal Insecticide, Fungicide and Rodenticide Act (FIFRA), originally enacted in 1947 and significantly amended in 1972 and 1975. The Environmental Protection Agency (EPA) is primarily responsible for regulating the use of pesticides. Pesticides are granted registration for use after submitting to the EPA results of thorough toxicological and environmental tests. The following sections will deal mainly with requirements for environmental testing of controlled-release pesticide formulations.

The test requirements are most detailed for the pesticide chemical itself (called the active ingredient) and less exhaustive for the dozens or hundreds of formulations containing the particular active ingredient. In the past, pesticides were formulated as dusts, wettable powders, or emulsifiable concentrates. Controlled-release pesticides represented a new concept in formulations, and the EPA technical staff did not have any precedents to guide them in their scientific review. The main question that arose was, Is the controlled-release formulation different from conventional products, or are different tests required to prove the environmental safety of the product?

The registration of Penncap-M by the EPA in 1974 represented a landmark, since it was the first major controlled-release pesticide approved by the government. The manufacturer of Penncap-M conducted a series of environmental and toxicological studies to convince the authorities that their product was "safe" for use. These studies were conducted on Penncap-M and the polymer, cross-linked nylon, which made up the formulation, and topics of study ranged from environmental fate, effect on fish and wildlife, to residues in food and feed crops. Comparative toxicological studies of metabolism in rodents and lactating animals were conducted and compared with results from methyl parathion EC formulations. In the final action, the EPA exempted the encapsulating material from the requirements of establishing a food tolerance.[4] It was only after extensive field use that Penncap-M appeared to be hazardous to pollinating insects, honeybees (see Section 3.2).

Controlled-release pesticide formulations must be treated as a special class of product, and certain test requirements must be met before governmental registration can be granted. This is not to suggest that these tests must be as extensive as those required for a new pesticide chemical, but that the tests may be uniquely designed to answer questions of hazard to beneficial nontarget organisms.

The requirements for registration of encapsulated pesticide formulations were recently discussed,[18,19] Studies on the environmental fate of the pesticide from an encapsulated product will be helpful in predicting potential environmental problems. These studies should cover both the active ingredient being released from the encapsulating material and the polymeric material itself. Comparative studies of the controlled-release product and another formulation (e.g, EC) will be helpful in contrasting the behavior of the active ingredient. Specific questions that must be resolved by these studies may include the following:

Does the polymeric material act as a soil conditioner?

What is the environmental persistence of the polymer?

What is the fate of the active ingredient in a conventional formulation and a controlled-release formulation?

What is the mechanism of the behavior of the chemical? Is it due to: microbial degradation? leaching in soil? hydrolysis? accumulation in aquatic organisms?

Other questions that must be addressed deal with the effect on aquatic and non-target organisms, especially beneficial insects like bees. These questions are:

What is the effect on fish and wildlife (comparison between controlled-release and the active ingredient itself)?
What is the effect on nontarget species? What is the effect of capsule size?
Is there a potential problem of insect resistance developing?

The Guidelines for Environmental Chemistry, Hazard Evaluation for Wildlife, Aquatic Organisms, and Nontarget Insects will be issued in 1982. These guidelines will then serve as guiding protocol for conducting tests on the environmental impact of controlled-release formulations of pesticides and the elastomeric material without the active ingredient. In many cases the manufacturer of the polymer may not necessarily be the pesticide manufacturer, has already conducted environmental tests on his products and may make these studies available to the pesticide manufacturer.

5. SUMMARY

A large number of diverse controlled-release pesticide formulations have been developed experimentally, but there are probably fewer than 10 commercially important products.[1] To facilitate registration, it will be necessary for manufacturers of polymers used in controlled-release products to develop basic information on the environmental impact of the polymeric matrix. From the scanty literature on this subject, it appears that most polymers used in controlled-release formulations are slowly degraded environmentally, and those that do degrade at a faster rate, form innocuous degradation products. Also it seems from analyzing the published data that most commonly used polymeric matrices are not hazardous to living organisms. Before a new controlled-release pesticide product can be used (registered) it must be ascertained that it does not cause environmental damage to nontarget organisms, especially to beneficial insects and fish. Specific guidelines for testing controlled-release pesticide products have not been established by the EPA, so each new product will have to be evaluated on its own merit. Optimistically, however, as our knowledge on the environmental fate and effect of controlled pesticides expands, guidelines can be formulated on the type of specific test needed to evaluate the environmental safety of these useful products.

REFERENCES

1. N. F. Cardarelli and K. E. Walker, *Development of Registration Criteria for Controlled Release Pesticide Formulations*, EPA-540/9-77-016, U.S. EPA, Washington, DC, 1978, 142 pp; for sale by NTIS, Springfield, VA, 22161.

2. H. Geldolf, "Resistance of Natural amd Synthetic Rubber Against Microbial Attack," *Tech. Semin. Ind. Rubber Manuf. Assoc.,* **1966**, 89–97; *Chem. Abstr.,* **70**, 12382 (1969).

3. Donald V. Rosato, in Dominick V. Rosato and R. T. Schwartz, Eds., *Environmental Effects on Polymeric Materials*, Vol. 1, Wiley-Interscience, New York, 1968, Ch. 8, p. 733.

4. *Code of Federal Regulations,* **40**, "Protection of the Environment," 1980, 180.1028.

5. *Code of Federal Regulations,* **40**, "Protection of the Environment," 1980, 180.1053.

6. W. M. Heap and S. H. Morrell, *J. Appl. Chem.,* **18**(7), 189 (1968).

7. *Code of Federal Regulations,* **40**, "Protection of the Environment," 180.1038.

8. *Code of Federal Regulations,* **40**, "Protection of the Environment," 1980, 180.1039.

9. F. L. Lyman, "Human Safety and Pesticide Use: The Role of Microencapsulation," in *Proc. Informal Conf. 1979 Annual Meet. Entomol. Soc. Am., Microencapsulated Insecticides: Agric. Benefits and Effects on Honey Bees,* Nov. 27, 1979. (Available from Pennwalt Corp., Fresno, CA, 93710.)

10. P. Mathiessen, "Some Effects of Slow-Release Bis(tri-*n*-butyl tin) Oxide on the Tropical Food Fish *Tilapia mossambica* Peters," in *Controlled Release Symp., Univ. Akron, OH, 1974,* Paper No. 25.

11. J. Selkirk, *J. Environ. Action Bull.,* **3**, Oct. 2, 1976, as cited by G. Zweig, Chapt. 4, in H. B. Scher, Ed., *Controlled Release Pesticides (ACS Symp. Ser. 53),* Am. Chem. Soc., Washington, DC, 1977, p. 37.

12. R. J. Barker, Y. Lehner, and M. E. Kunzmann, *Z. Naturforsch.,* **34**(c), 153 (1979).

13. M. Burgett and G. Fisher, *Am. Bee J.,* **117**, 626 (1977).

14. J. R. Lowell, Jr., *"The Benefits of Microencapsulation, as Demonstrated in Penncap-M* Insecticide. A Review of Five Years Commercial Experience,"* Report by Pennwalt Corp., Agric. Chem. Div., Fresno, CA, July 11, 1979, 44 pp.

15. N. F. Cardarelli and W. Evans, "Chemodynamics and Environmental Toxicology of Controlled Release Organotin Molluscicides," in R. Baker, Ed., *Controlled Release of Bioactive Materials*, Academic, New York, 1980, pp. 357–385.

16. N. Lahav and D. Tropp, *Soil Sci.,* **130**(3), 151 (1980).

17. G. A. Janes and S. Z. Mansdorf, *Abstr., 7th Int. Symp. Controlled Release of Bioactive Materials, Ft. Lauderdale, FL, July 28–30, 1980,* pp. 176–179.

18. G. Zweig, "Current Views on the Requirements for the Registration of Encapsulated Pesticide Formulations," in *Proc. 5th Int. Symp. Controlled Release Bioactive Materials,* Nat. Bur. Standards, Gaithersburg, MD, 1978, pp. 150–155.

19. H. S. Harrison, "Pesticide Regulation and Controlled Release Products," presented at *7th Int. Symp. Controlled Release Bioactive Materials,* Ft. Lauderdale, FL, 1980.

CHAPTER EIGHT

Chronic vs. Acute Intoxication

N. F. CARDARELLI
C. M. RADICK
Environmental Management Laboratory
University of Akron
Akron, Ohio

CONTENTS

1. INTRODUCTION

Conventional pesticide treatment methodology consists of the application of a relatively large amount of the chemical agent to the target habitat. Normal dosages, which vary from less than 1 to over 30 ppm, are "massive" in the sense that the amount applied is always far in excess of the amount required if we note the lethal dose per individual target multiplied by the number of target organisms to be destroyed. The quantity used depends upon the agent of choice, the target species, and the nature of the environment to be intoxicated. This chapter considers only aquatic environments and the application of herbicides, insect larvicides, and molluscicides.

Of necessity, large quantities of the given pest control agent are used in order to overcome dispersal through water flow, which results in a rapid drop in concentration, and detoxication processes arising from chemical reaction with dissolved or suspended matter, hydrolysis, solar radiation, and so on. The effective half-life of a given pesticide molecule in a natural water body may be only a few hours. During this restricted exposure time, the target species must absorb a lethal dose.

Controlled-release pesticides consist of a dispensing unit such as a granule, pellet, or strand that is composed of an agent bound by a polymeric matrix. In some instances the agent is monolithically incorporated in an elastomeric or thermoplastic material, uniformly dispersed throughout the free volume of the polymer matrix. The chemical agent is released through a diffusion-dissolution mechanism that relies on interstitial movement through solution pressure; leaching, wherein the presence of a pore structure is essential; or the chemical cleavage of a pendent toxic moiety, for instance, through hydrolysis. In all cases emission is slow and continuous, and depletion of a given dispenser may require up to several years. If it were necessary to maintain a high toxicant concentration in the water course treated, say 5 ppm or greater, the volume of controlled-release material necessary would be prohibitive. Consequently, one of the problems addressed early in the development of controlled-release pesticides was to determine the application dosages that would expose the target species population to the agent continuously rather than periodically as with conventional technology.

2. CONCENTRATION × TIME RELATIONSHIP

It has long been presumed that the concentration × time, or CT, relationship is a valid method of determining pesticide application dosage. That is, if target exposure at 2 ppm for 6 h provides a lethal dose for some segment of the population, then 1 ppm for 12 h or 4 ppm for 3 h should be equally effective.[1] Universal acceptance of this doctrine is convincing evidence of at least an approximate reliability. Obviously there are upper limits based on respective kill mechanisms. Snails exposed to 1 ppm copper ion concentration under laboratory conditions succumb within 6 h to an acute intoxication. However, at 100 ppm copper ion concentration, the same species, *Biompharlaria glabrata*, do not die in 0.06 h (3.6 min). Sufficient time is necessary for the copper ion to move into the molluscan target, be transported through internal tissue, and accumulate in a lethal dosage at sites of cidal activity.

The critical question regarding controlled release is whether the *CT* relation holds at dosage levels practical to the emission of an agent from a polymeric matrix. If the necessary dosage is 2 ppm for a given plant species to succumb to a given herbicide, and the practical emission rate will provide only 0.01 ppm per day, then 200 days would be required for control. Obviously such a formulation would have severely limited value, if any.

Fortunately the *CT* relationship is not valid at ultralow dosage levels. The following sections of this chapter will provide supporting data for that statement.

3. ACUTE TOXICITY: HERBICIDES

The biological activity of a herbicide depends on many complex factors. Although no single way of expressing potential effect is completely reliable, the most rapid and convenient indicator for most chemicals is acute toxicity (the single dose necessary to produce death). The usual way of expressing acute toxicity is by means of an LD_{50} (median lethal dosage) value. The LD_{50} is a statistical estimate of the dosage that would be lethal to 50% of a large population of the test species. Although LD_{50} values give no information on the dosage that would be lethal to every individual of the species, or to treatment given in some other way than in the test, the LD_{50} value, within its confidence limits, is still probably the most convenient and reliable means available for comparing the inherent toxicity of chemicals. There is enough similarity between cases to establish rules of thumb relating lethal levels for a species in laboratory toxicity tests to potential effects from field exposure. One method is to predict the field effects of a new chemical by comparing its LD_{50} for the species of interest with the LD_{50} of a chemical whose field effects are known. Thus, if herbicide X has an LD_{50} of 10 mg/kg and is known to kill plants in the field when applied at 2.24 kg/ha, herbicide Y with an LD_{50} of 5 mg/kg might be expected to kill plants at 1.12 kg/ha.[2]

4. CHRONICITY PHENOMENON: HERBICIDES

In order to determine the dosage levels necessary to destroy various aquatic weeds through herbicide emission from a polymeric matrix, a number of laboratory experiments were performed. During the course of this activity it was noted that the *CT* presumption was not valid when a given target was exposed to continuous herbicide stress over a long period of time at very low concentrations.

In one series of experiments, several select aquatic weeds were grown in 4-liter aquaria under laboratory conditions. Temperatures ranged from 60 to 75°F depending upon the species examined. Gro-lux lighting was used at levels optimum to plant growth at a 12 h on–12 h off cycle. After stabilization, a given herbicide was introduced at freshly prepared known concentrations. In one group of experiments, a given low or ultralow concentration was introduced daily. In a second group of experiments, about 90–95% of the exposed water was siphoned off and preconditioned fresh water was added prior to application of the known herbicide dosage. That is, in the first group, herbicide concentration gradually increased to some equilibrium limit where degradative forces removed an amount of agent per day equivalent to the newly added amount. In the second instance, dosage was held nearly constant over the test period.

In all experiments approximately 250 g of soil was added to each container to serve as a source of nutrition. The soil mixture was 50% high organic peat moss and 50% sand. Plants that normally root were planted in the soil mixture, whereas aquatic species that normally float, such as *Lemna minor,* were floated over the soil. Larger plants, *Eichonia crassipes* (water hyacinth), *Polygonum* (smartweed), and *Alternanthera philoxeroides* (alligator weed) were tested in 20–40-liter aquaria containing 2000 or 5000 g of the soil mixture.

Observations were made daily, plant conditions being noted according to the presence of lifeless tissue darkening, loss of leaves, and stem breakage. A 10-point scale was used wherein 10 signified a healthy growing plant and zero a plant devoid of leaves and with general decay of other structure. Obviously, the point of lethality for each plant could not be checked under the conditions of the experiment. In ancillary recovery experiments, exposed plants having an index of 3 or less never recovered. From the practical standpoint, a plant devoid of leaves and stems, or a floating plant that has decayed and sunk, are no longer of a pestiferous nature, and control has been achieved.

All tests were performed in replicates of five or more test aquaria. Usually three plants per aquarium were used in the 4-liter containers and five plants each in the 20–40-liter containers. Various runs were repeated for verification. Table 1 provides data on the herbicidal agents used, and Table 2 lists the plant species evaluated.

Tables 3 and 4 depict plant degradation upon exposure to an accumulating daily dosage of a given agent. The criterion used is the lethal time to a given population mortality; that is, an LT_{90} of 21 days indicates that 90% of the test population had a zero rating on the 21st post-exposure day. Tables 5 and 6 depict experimental runs where the dosage was held fairly constant through daily water change.

TABLE 1. Herbicides Used

Trade or Laboratory Designation	Chemical Name	Type
Diquat[TM] (Chevron Chemical Co.)	6,7-Dihydrodipyrido(1,2-a:2',1'-c) pyrazidiinium dibromide	Technical 97%
2,4-D BEE (Dow Chemical Co.)	Butoxyethanol ester of 2,4-dichlorophenoxyacetic acid	Technical 97%
Fenac[TM] (Amchem Products, Inc.)	2,3,6-Trichlorophenylacetic acid	Technical 100%
Silvex[TM] (Dow Chemical Co.)	2-(2,4,5-Trichlorophenoxy) propionic acid	Technical 100%
2,4-D acid (Dow Chemical Co.)	2,4-dichlorophenoxyacetic acid	Technical 96%
2,4-D butyl ester (Dow Chemical Co.)	Butyl ester of 2,4-dichlorophenoxyacetic acid	Technical 96%
Endothall (Pennwalt Corp.)	7-Oxabicyclo (2,2,1)heptane-2,3-dicarboxylic acid	Technical 96%
Fenuron (E.I. DuPont de Nemours & Co.)	1,1-Dimethyl-3-phenylurea	Technical 98%
2,4-D oleylamine (Dow Chemical Co.)	Oleylamine salt of 2,4-dichlorophenoxyacetic acid	Technical 94%
2,4-D amine (Dow Chemical Co.)	Dimethylamine salt of 2,4-dichlorophenoxyacetic acid	Technical 97%
Acrolein (Shell Chemical Co.)	2-Propeno	Chemically Pure 100%
Hydrothol (Pennwalt Corp.)	Sodium salt of Endothall	Technical 96%
Dichlobenil (Thompson-Hayward Chemical Co.)	2,6-Dichlorobenzonitrile	Technical 99%

The 56+ designation indicates that the exposed population did not reach the noted mortality level over the 56-day duration of the experiment. Control deaths were 15% or less for all runs except that of the 2,4-D butyl ester, where a 50% mortality level was reached by day 56 in the 0.1 ppm/day group.

Table 7 summarizes the various evaluations as to the presence or absence of the chronicity phenomenon as it applies to several aquatic plant species. Data are taken from published and unpublished sources.

In summary, laboratory studies of phytotoxicants indicate that the following generalized situation exists. In the range of approximately 0.5–20+ ppm, the CT relationship holds and plants are destroyed within a corresponding exposure

TABLE 2. Plant Species Evaluated

Common Name	Biological Name	Type
Water hyacinth	*Eichornia crassipes*	Floating
Alligator weed	*Alternanthera philoxeroides*	Rooted
Elodea	*Elodea canadensis*	May root
Eurasian watermilfoil	*Myriophyllum spicatum*	May root
Vallisneria	*Vallisneria americana*	Rooted
Cabomba	*Cabomba caroliniana*	May room
Water lettuce	*Pistia stratiotes*	Floating
Southern naiad	*Najas guadalupensis*	May root
Coontail	*Ceratophyllum demersum*	Rooted reed
Smartweed	*Polygonum*	Rooted reed
Duckweed	*Lemna minor*	Floating

period. In the 0.001-1 ppm range, the *CT* relationship does not hold and mortality is induced within an exposure period varying from 5 or 6 days to 60 or more days. Below a limiting threshold, which varies widely with different species from 0.005 to 0.0005 ppm, exposures of 60-90 days show no appreciable gross effects.[8]

5. HYPOTHESIS CONCERNING THE CHRONICITY PHENOMENON IN REGARD TO PLANTS

Ultralow herbicide concentrations have been observed to slowly affect 12 major aquatic weed species, such exposures terminating in nearly 100% mortality for two to eight species. A chronic intoxication mechanism is obviously involved, and the question arises as to the nature of the chronic intoxication mechanism. As long ago as 1960, Shaw et al. inferred the existence what is now called the chronicity phenomenon.[9] While relatively little work on aquatic herbicidal mechanisms has been performed, a great deal of information is available concerning terrestrial plants. In all cases the phytocidal molecule, or a molecule that will react with enzymes or other factors within the plant to become phytocidal, must contact the target, penetrate, and reach the site of cidal activity. In water and soil, various detoxification factors exist so that the quantity of herbicide moving from granule, pellet, powder, or emulsified droplet must be sufficient to overcome such natural processes, with the concomitant loss of a portion of the total

TABLE 3. LT Values for *Myriophyllum spicatum* (Eurasian Watermilfoil) Exposed to Various Herbicides via Daily Dosing[a]

Herbicide	Concentration (ppm/day)	Days to a Given Mortality			
		LT_{25}	LT_{50}	LT_{90}	LT_{100}
Diquat	1.0	5	9	13	14
	0.1	6	9	13	14
	0.01	7	10	14	19
	0.001	13	19	24	38
2,4-D BEE	1.0	10	10	11	15
	0.1	10	11	15	17
	0.01	10	11	14	18
	0.001	10	11	15	22
Fenac	1.0	10	14	18	21
	0.1	14	31	42	43
	0.01	21	30	42	48
	0.001	23	56+	56+	56+
Silvex	1.0	5	8	13	14
	0.1	11	17	23	43
	0.01	11	14	20	56+
	0.001	22	56+	56+	56+
2,4-D acid	1.0	9	12	18	20
	0.1	14	17	56+	56+
	0.01	26	34	56+	56+
	0.001	56+	56+	56+	56+
2,4-D butyl ester	1.0	3	5	14	20
	0.1	8	11	14	22
	0.01	12	15	23	24

[a] 5 replicates × 3 plants per replicate.

numbers of molecules added, and contact the plant with a certain minimum amount of toxicant. Penetration is a slow process, and natural degradation occurs as the molecules are in surface contact prior to actual entry.[10,11] Penetration rate depends upon pH, light, and other factors.[11,12]

Contact herbicides destroy plant tissue at or near the point of application through denaturation of the plant cells, uncoupling of respiratory oxidation from phosphorylation, or other mechanisms. This type of destruction may be

TABLE 4. LT Values for *Vallisneria americana* Exposed to Various Herbicides via Daily Dosing[a]

Herbicides	Concentration (ppm/day)	Days to a Given Mortality			
		LT_{25}	LT_{50}	LT_{90}	LT_{100}
Diquat	1.0	4	5	5	6
	0.1	5	5	7	12
	0.01	8	11	18	31
	0.001	18	23	46	60
2,4-D BEE	1.0	9	11	15	18
	0.1	11	17	31	45
	0.01	19	23	40	51
	0.001	22	26	46	56
Fenac	1.0	12	13	17	34
	0.1	12	14	22	35
	0.01	20	31	43	56
	0.001	56+	–	–	–
Silvex	1.0	15	23	27	34
	0.1	27	39	49	60
	0.01	51	55	59	61
	0.001	49	56+	–	–
2,4-D acid	1.0	9	10	12	12
	0.1	24	29	33	59
	0.01	29	34	42	60+
	0.001	28	56+	–	–

[a] 5 replicates $\times$ 3 plants per replicate.

common to a number of aquatic herbicides. Materials examined in the light of the chronicity effect, with the possible exception of acrolein, appear to be translocated through plant tissue to a cidal point.

In terrestrial plants, the control agent must, in many cases, enter the stomata and/or penetrate the cuticle, migrate across the mesophyll into the phloem, and move with the assimilate stream to the meristematic regions of the roots or possibly to other cidal sites. 2,4-D penetrates the cuticle and enters the aqueous phase of the mesophyll. There the pH is 5–5.5, which should cause dissociation and molecular diffusion into the cytoplasm and hence to the phloem. Movement is probably along the symplast to the cidal area.[13] Effectiveness depends upon the amount of herbicide reaching a site of phytotoxic action.

TABLE 5. LT Values for *Myriophyllum spicatum* Exposed to Various Herbicides at a Constant Dosage[a]

Herbicide	Concentration (ppm/day)	Days to a Given Mortality			
		LT_{25}	LT_{50}	LT_{90}	LT_{100}
Diquat	1.0	6	9	11	11
	0.1	6	9	13	19
	0.01	6	9	13	16
	0.001	15	19	25	32
2,4-D BEE	1.0	7	8	11	14
	0.1	7	9	14	22
	0.01	7	11	22	27
	0.001	12	20	32	56+
Fenac	1.0	12	19	30	36
	0.1	24	28	37	39
	0.01	20	56+	—	—
	0.001	22	56+	—	—
Silvex	1.0	6	8	12	15
	0.1	13	14	18	21
	0.01	18	21	23	28
	0.001	22	24	27	32
2,4-D acid	1.0	9	12	18	20
	0.1	14	17	21	29
	0.01	26	34	35	39
	0.001	29	40	46	56+
2,4-D butyl ester	1.0	4	5	7	7
	0.1	5	6	8	10
	0.01	6	7	9	10
	0.001	15	34	60+	—
2,4-D oleylamine ester	1.0	4	5	6	7
	0.1	7	9	11	12
	0.01	9	12	16	17
	0.001	19	28	47	60+

[a] 5 replicates × 3 plants per replicate.

TABLE 6. LT Values for *Vallisneria americana* Exposed to Various Herbicides at a Constant Dosage[a]

Herbicide	Concentration (ppm/day)	Days to a Given Mortality			
		LT_{25}	LT_{50}	LT_{90}	LT_{100}
Diquat	1.0	4	4	4	4
	0.1	4	4	6	20
	0.01	10	16	50	64
	0.001	15	27	56	70+
2,4-D BEE	1.0	8	24	37	42
	0.1	10	16	38	45
	0.01	12	25	39	51
	0.001	17	27	38	51
Fenac	1.0	10	10	16	19
	0.1	10	12	19	23
	0.01	10	19	27	42
	0.001	16	20	43	55
Silvex	1.0	11	19	29	41
	0.1	13	20	23	44
	0.01	16	19	24	51
	0.001	16	19	29	58
2,4-D	1.0	14	16	17	22
	0.1	21	24	27	29
	0.01	22	31	35	49
	0.001	24	32	49	60+

[a] 5 replicates × 3 plants per replicate.

The toxic molecule must follow a critical path, and each element of that path poses a physical or biochemical barrier. For instance, considering a terrestrial plant, the presence of 2,4-D molecules on the foliage triggers a physiological response culminating in the closure of the stomata, thus erecting a physical block to herbicide penetration. The cuticle acts as a barrier per se, and penetration requires that the agent have a specific chemical structure. Once through the cuticle the toxic agent must overcome hydrophilic and nonpolar constituents of cells. If the mesophyll is penetrated, then various barriers to translocation are encountered. Metabolic reactions involving one or more plant enzymes, or nonenzymatic activity, can detoxify the herbicide. Metabolic reactions fall into several categories: oxidation, mainly oxidative dealkylation or less frequently demethyla-

TABLE 7. Presence (+) or Absence (−) of the Chronicity Phenomenon—for Various Plant–Herbicide Combinations[2-7]

	2,4-D BEE	Diquat	2,4-D Butyl Ester	2,4-D Amine	2,4-D Oleylamine	Silvex	Fenac	2,4-D Acid	Endothall	Acrolein	Hydrothol	Dichlobenil	Fenuron
Water hyacinth	+	+	−	−	+	−	−	*	−	−	−	+	*
Watermilfoil	+	+	+	+	+	+	+	+	−	*	−	+	*
Water lettuce	+	+	*	*	*	+	+	+	*	−	*	*	*
Vallisneria	+	+	+	*	*	+	+	*	*	*	*	+	+
Elodea	+	+	+	+	*	+	+	+	−	−	−	+	+
Cabomba	+	+	*	*	*	−	+	+	+	*	*	+	+
Duckweed	+	+	+	*	*	+	*	*	+	−	*	*	*
Southern naiad	*[a]	+	*	*	*	+	+	*	+	*	*	+	*
Alligator weed	+	+	*	*	*	+	*	*	*	−	*	*	*
Smartweed	+	+	*	*	*	+	*	*	*	*	*	*	*

[a] Asterisks indicate plant–herbicide combinations that were not evaluated.

205

tion; hydroxylation; synthesis of conjugates wherein glycosides predominate; and hydrolysis or reduction.[14]

2,4-D may also enter the plant through root hairs and cortex cells and migrate via the symplast to the stele and across the stele to the apoplast. Movement upward probably occurs in the transpiration stream, although lateral accumulation can take place in the xylem.

Once the phytocide reaches the cidal point, various disruptions of vital physiological factors occur, terminating in mortality, if a sufficient number of herbicide molecules are present. Critical interference with respiration, phosphatase activity, potassium metabolism, carbohydrate utilization, and so on, or blockage of photosynthesis can be the underlying mechanism.

5.1. Plant Response to Stress

It is well known that a plant exposed to a physical or chemical stress will react to neutralize the effect of that stress. Total response is dependent upon the degree of stress. If the stress exceeds a certain limit, protective mechanisms are overwhelmed, and serious or terminal injury results. Below this limit the induced physiological strain is elastic, and once the stress is removed the protective devices disappear and the plant returns to its normal state. A lower threshold must also exist. The situation must often arise where the stress elements are present but in such slight degree as to invoke no response. Further analysis would therefore indicate that another threshold must exist.

This view can be summarized as follows:

Zone	Stress Factors	Results	Condition	Limit
I	Present	Injury/mortality	Stress overcomes resistance to that stress	Plastic
II	Present	Recovery	Stress resistance is equal to stress applied	Elastic threshold
III	Present	None (?)	No response	Null Threshold
IV	Absent	None	No response	

The existence of an injury/mortality zone separated by the plastic threshold and the presence of an elastic threshold are thoroughly substantiated by a large body of relevant literature.[15] The existence of a null threshold can be challenged.

Consider a herbicide as the environmental factor underlying the stress. Consider 2,4-D action against a given target aquatic plant, for instance *Hydrilla*. Under certain conditions of water quality and exposure duration, say 12 h, 1 ppm will destroy this plant or, to be statistically precise, the LD_{99} is 1 ppm. The LD_5 is perhaps 0.4 ppm. Thus we can define the plastic limit in terms of herbicide-induced stress as a function of concentration as the LD_5-LD_{99} range, or 0.4-1.0 ppm. (The LD_0 value, while experimentally determinable, is not mathematically sound.) Below 0.4 ppm, plant response is observed, epinasty for instance, but once the 2,4-D exposure ceases, plant recovery is noted.

If a statistically relevant number of plants are exposed to decreasing 2,4-D concentrations for the *short* exposure period of 12 h, eventually a point *must* be reached at which no response is detectable by any known analytical method. This can be defined as the elastic threshold. Suppose that this value is 0.08 ppm.

Zones I and II having been adequately described for the purposes of this discussion, it is appropriate to next consider zone IV by asking Is there a herbicide dose so small that it cannot affect the living plant cell? Stochastic considerations advanced by Dinman[16] and based upon the work of Hutchinson[17] would indicate that somewhere around 10^4 molecules of a toxic agent are minimal for cell response. Simple calculation would thus indicate that 10^4 molecules/plant cell is equivalent to a toxicant concentration of no more than 10^{-7} ppm within the cell. The environmental dosage that would provide this cellular concentration is unknown, but assuming a penetration loss of $\frac{1}{1000}$ (only one toxicant molecule reaches the cell for each 1000 added to the watercourse), then the null threshold is 0.0001 ppm.

Zone III is that grey area that, in the example, lies between the null threshold and the elastic threshold, or from 0.0001 to 0.08 ppm. Since the chronicity phenomenon is manifest between 0.001 and 0.1 ppm, it is reasonable to term zone III the chronicity zone.

The values given as defining zone III are obviously open to serious question, and should be. The critical choice lies in either accepting the existence of zone III or advocating that the elastic threshold and the null threshold are identical. To assume the latter, one must believe that the biological effect of *one* rather small toxic molecule on a plant cell composed of billions of molecules with quadrillions of reactive sites is the same as the effect of several billion toxicant molecules on the same cell.

Plants respond to external stimuli. The mechanism of detection of the stimulus and corresponding response is not well elucidated. Plants may possess a rudimentary nervous system, and some supporting data are available to that effect.[18] No matter what the nature of the external stimuli, electromagnetic radiation, moisture, pressure, or phytotoxic substances, a threshold must exist between that amount of stimulus that elicits response and a lesser amount that does not.

The following hypothesis is proposed: A herbicide concentration range exists wherein exposed plant life does not respond by erecting penetration barriers but free entry of the phytotoxic molecules leads to their accumulation at the cidal location, and with prolonged exposure a terminal chronic intoxication is induced.

In order to further elucidate this hypothesis, several experiments were performed. *Vallisneria americana* and *Elodea canadensis* were exposed to both a high (acute) and low (chronic) dosage of 2,4-D BEE and copper ion from copper sulfate pentahydrate. Table 8 depicts the LT_{100} observed. Plant leaf sections were removed at 15 min, 6 h, 24 h, and 120 h post-exposure and histologically examined for physiological change. Specimens were stained with hematoxylin and saffron fast green. The following were noted:

1. *Elodea* exposed to 2,4-D BEE at 4 ppm showed immediate foliar damage with cell-wall rupture and the concealing of protoplasm. By 6 h post-treatment, approximately 20% of the dermal cells showed pathogenic changes. However, this trend does not continue to mortality. Exposure at the chronic dosage level caused very little observable changes in tissue for the first 24 h exposure.

2. Copper reaction with leaf tissue was much more rapid, with approximately 80% of the dermal cells showing gross change within 15 min and nearly all cells affected within 6 h at the higher concentration. Chronic dosage effects were not noted until 24 h post-exposure. After 7 days, nearly all cells show complete destruction at both dosage regimes.

3. *Vallisneria* exposures to the two phytocides showed cell-wall collapse and fusion or congealing of contents similar to that observed with *Elodea*. Approximation of the number of cells grossly affected provides a measure of the rate of pathogenicity.

TABLE 8. **Plant Mortality: Conventional and Ultralow Herbicide Concentration Exposures**

Toxicant	Concentration	Lethal Time (LT_{100} (days)	
		Vallisneria	*Elodea*
2,4-D BEE	4 ppm[a]	4	6
	0.01 ppm/day[b]	9	14
Cu^{2+}	20 ppm[a]	5	3
	0.1 ppm/day[b]	13	15
Control	0	30+	30+

[a]Applied as a one-time on day zero.

[b]Applied daily to test aquaria.

Table 9 provides data on the dermal cell effect as a function of time.

Since foliar tissue damage occurs at both acute and chronic dosage levels, it is believed that the critical cidal activity takes place within the plant and destruction does not arise from contact alone. Assuming that the first line of defense against toxicant entry lies with the dermal tissue, then the lack of early cell stress may not trigger protective reactions at the chronic dosage level.

Janes has noted that water plants, fish, and snails exposed to copper ion from copper sulfate similarly exhibit a chronic intoxication effect at dosages far lower than that expected from adherence to the *CT* principle.[19]

Analysis of the lethal time ratios for aquatic plant destruction at an accumulating dosage is shown in Table 10. It is apparent that the *CT* relationship is inoperative.

TABLE 9. Cell Destruction: Vallisneria Exposure to 2,4-D BEE and Cooper Ion

| Time (Post-treatment) | Approximate percent of affected dermal cells | | | | |
| | 2,4-D BEE | | CU^{2+} | | |
	Acute	Chronic	Acute	Chronic	Control
15 min	0	0	20	0	0
6 h	60	0	80	0	0
24 h	100	0	100	40	0
7 d	—	40	—	80	0

TABLE 10. LT Ratios LT_{90} (1 ppm)/LT_{90} (0.01 ppm) for Various Herbicide-Plant Combinations

Herbicide	Vallisneria	Cabomba	Eurasian Watermilfoid	Elodea	Southern Naiad
2,4-D BEE	2.7	1.1	1.3	1.8	—
Diquat	3.6	1.3	1.3	1.8	1.7
Silvex	1.8	—	1.5	3.3	—
Fenac	1.6	2.5	2.3	3.7	1.6
2,4-D acid	—	—	—	1.4	—
Endothall	—	1.1	—	—	1.8
Fenuron	—	—	—	1.8	—

Chronic herbicide dosages for 30- and 60-day aquatic plant mortality have been calculated. Results are depicted in Table 11. Variations exist between species due to specific tolerance.

6. MOLLUSCICIDES AT ULTRALOW CONCENTRATIONS

Molluscicides are used in much of the tropical world to control the snail hosts of schistosomiasis and other parasitic diseases.

All commonly used molluscicides are nonpersistent. Rapid detoxification occurs through chemical interaction with dissolved minerals and gases, organic absorption, inorganic absorption, and absorption by suspended matter and bottom soil; biodegradation through bacterial, algea, and fungal attack; and insolation. Laboratory results concerning dosages translate poorly to field conditions. Usually laboratory dosages are multiplied by a factor of 5–25 before field application is attempted.[20] Conventional treatment results in acute intoxication and rapid mortality of the target, usually within 24 h.

Controlled-release molluscicides based upon trialkylorganotins and copper sulfate incorporated in an elastomeric or plastic base have been developed.[5,21,22] Table 12 provides a generalized comparison between a single molluscicide dose

TABLE 11. Chronic Herbicide Dosages for LT_{90} —30 Days and LT_{90} —60 days[5]

		H	Dosage (ppm/day)		
Herbicide	Lethal Dose	Cabomba	Vallisneria	Watermilfoil	Elodea
2,4-D acid	LT_{90} —30 days	0.1	0.09	0.008	0.0009
	LT_{90} —60 days	0.1	0.05	0.001	0.0005
Silvex	LT_{90} —30 days	0.13	0.9	0.007	0.05
	LT_{90} —60 days	0.08	0.09	0.002	0.008
Diquat	LT_{90} —30 days	0.08	0.006	0.0008	0.03
	LT_{90} —60 days	0.015	0.0004	0.0005	0.0006
2,4-D BEE	LT_{90} —30 days	0.2	0.006	0.004	0.0006
	LT_{90} —60 days	0.008	0.002	0.003	0.0003
Fenac	LT_{90} —30 days	0.0008	0.07	0.4	0.5
	LT_{90} —60 days	0.0004	0.02	0.008	0.08
Endothall	LT_{90} —30 days	0.2	0.09	0.09	—
	LT_{90} —60 days	0.1	0.06	0.06	—
Fenuron	LT_{90} —30 days	0.2	0.04	—	0.08
	LT_{90} —60 days	0.09	0.01	—	0.007

TABLE 12. Conventional vs. Controlled Release: 24-h LD_{99} Values (Laboratory Conditions)

Material	Conventional Dosage (ppm)	Controlled-Release Dosage (ppm/day)
Copper sulfate	2	0.03
Niclosamide	0.2–0.4	0.005
Trifenmorph	1	–
Tributyltin oxide	0.3	0.006
Tributyltin fluoride	0.2	0.002

administered in the conventional sense and the observed continuous dosage necessary to achieve practical snail mortality.

In order to transcend the gap between relatively high molluscicide dosages required for snail destruction under field conditions and LD values measured in the laboratory, microecological systems were prepared in which various agents were evaluated. One-gallon wide-mouth glass jars were used as the test containers. A soil mixture comprising 400 cm^3 sterilized sand, 400 cm^3 top soil, 400 cm^3 peat moss, and 50 g limestone chips was placed in each aquarium, 3000 cm^3 mineral spring water was added, and aeration was begun. After several days, a 5-8-g sprig of *Elodea canadensis* was added. These usually rooted within a few days. After one week of conditioning, water quality was checked. Ten mature (7-9-mm shell diameter) noninfected *B. glabrata* snails and ten adult *L. reticulatus* guppies (0.7 in. or greater body length) were added from stock aquaria. After an additional 2 weeks of conditioning with daily observation for fish or snail deaths, intoxication was initiated.

Copper sulfate pentahydrate (C.P.), niclosamide (96% tech.), and TBTO (96% tech.) water solutions were prepared. Ethanol was used to solubilize niclosamide and bis(tri-*n*-butyltin) oxide (TBTO). The maximum amount of ethanol added to the test aquaria never exceeded 0.008%/day. Past experience has shown that 0.05% ethanol has no effect on *B. glabrata, L. reticulatus,* or *E. canadensis.*

A measured toxicant dose was added daily to each aquarium. Dosage was based on a range from 100% fish and snail mortality to 0% fish and snail mortality.

Snails were fed lettuce *ad libidum.* Fish food (Tetramin) was supplied every other day. Aeration was continuous. Water lost through evaporation was replaced daily. Temperature varied from 70 to 74°F.

Mortality observations were made and recorded daily. Snail death was determined by microscopical examination for heart beat. Animals were removed within 24 h of death. Appropriate controls were established.

Toxicant exposure was continuous for 120 days.

Table 13 depicts the snail mortality observed for each molluscicide. The given agent was added daily so that accumulation occurred. The active life of each agent is not known under these conditions. However, in-house measurements indicate a $T_{1/2}$ of 16–57 days, depending upon water quality, for TBTO; and less than 1 day for $CuSO_4$.

7. CONTROLLED-RELEASE MOLLUSCICIDES

Bis(tri-*n*-butyltin) oxide and tributyltin fluoride were incorporated in various elastomeric polymers and through proper formulation were released at a slow continuous rate in water. Long-term antifouling and molluscicidal activities were noted.[5,23,24] Long-term molluscicidal activity was first observed at the University of Akron and later confirmed under laboratory conditions by Berrios-Duran and Ritchie.[25] Hopf and Goll[26] noted continuous activity over a 209-day period. Laboratory pond experiments in Tanzania also confirmed long-term activity.[27]

TABLE 13. Long-Term Snail Exposure to Several Molluscicides at Various Concentrations

Agent	Dosage (ppm/day)	Cumulative Mortality (%)			
		5 days	30 days	60 days	120 days
Copper ion	0.65	100	—	—	—
	0.13	22	86	100	—
	0.065	12	42	94	100
	0.013	2	18	58	74
	0.0026	0	2	28	38
	0.00	0	4	4	12
TBTO	0.035	58	100	—	—
	0.007	6	80	96	100
	0.0014	0	4	14	36
	0.0007	0	6	20	26
	0.00	2	4	4	4
Niclosamide	1.00	100	—	—	—
	0.50	100	—	—	—
	0.10	62	100	—	—
	0.05	0	32	94	100
	0.00	0	4	4	4

In each case the total water concentration was far lower than necessary to provide a 24-h LD_{90} and kill time required 4 or more days. Difficulties in measuring exact aqueous agent concentrations prevented extracting knowledge of agent release. However, dividing the amount of organotin known to be in the elastomer by the total number of days in which release had occurred as noted by 100% snail mortality indicated that the water concentration could not be in excess of 5-15 parts per trillion. Thus continuous exposure at ultralow concentrations provided snail kill at levels 1/30 to 1/100 of that necessary for acute intoxication.

In laboratory and small-scale field evaluations it was reported that in Brazil,[28-31] a 3-12 ppb/day organotin loss controlled snail population over a 9-month test period, and in Tanzania,[32] St. Lucia,[33] Iran,[34] and the Phillippines,[35] downstream transport was found to occur and snails were destroyed 150 m away from the dispensing pellet. In Rhodesia,[36-38] large ponds and the margins of lakes were effectively treated.

8. CHRONIC INTOXICATION OF SNAILS

Laboratory studies have demonstrated several distinct and unanticipated advantages arising from the use of controlled-release organotins. The value of continuous application of a molluscicidal agent at low dosage, such as in the "chemical barrier" concept, has been noted by Malek[39] and others. It had long been assumed that Haber's rule ($Ct = C't'$ for a given LD_{50}) was valid and would apply for continuous dosages of molluscicides. Consequently, it was believed possible to calculate minimal concentrations for prolonged exposure periods that would provide adequate snail control. Surprisingly, computation indicated dosage needs far in excess of those observed efficacious in the laboratory. Fenwick[27] anticipated the chronicity phenomenon in snails when he noted that increasing organotin concentrations did not proportionately reduce lethal time.

Fenwick examined code 633B containing both 3% TBTO and 8% Bayluscide (ethanolamine salt of niclosamide) under flowing-water conditions in the laboratory. He observed that *Biomphalaria pfeifferi* were destroyed in each of over 82 consecutive beaker tests.[27] It was suspected, from the work of Hopf and Goll, that the Bayluscide in 633B was not released and that the toxic element was the TBTO.[40] Fenwick noted in a concentration series that 3.3% of the maximum dosage tested killed in only twice the exposure time—evidence for the presence of a chronic intoxication mechanism. In laboratory ponds with a flow rate of 250 ml/min, 100% mortality was achieved over a 96-216-h period depending upon dosage.[27] It was further established that an initially high loss rate is experienced for the first few days of immersion. This is in accordance with the experience with antifouling rubber, which requires 2-7 days immersion before a steady-state release is attained.

A number of investigators have noted that continuous snail exposure to organotin-emitting devices results in a chronic intoxication syndrome and delayed, though inevitable, mortality. Whether this is due to an accumulation of the toxic moiety in the snail as suggested by Gilbert et al[30] is under investigation. The *CT* relationship is not operable when the duration of exposure is extended and the organotin content is below about 0.01 ppm.

Hopf and Goll examined controlled-release elastomeric formulations containing both niclosamide and TBTO. In repeated exposures against adult *Biomphalaria glabrata* over a 209-day period, it was reported that the TBTO-containing materials were superior in activity to the TBTS [bis(tri-*n*-butylin) sulfide] and TBTR (tributyltin resinate) samples, although all organotin materials continued to cause mortality over the test period.[26] Formulations submerged beneath a layer of mud likewise released sufficient toxicant into the supernatant, after saturation of the mud, to provide 100% snail mortality.[26] Importantly, it has been shown that the elastomer used plays a crucial role in the rate of release.[40] An estimated biological life of more than 26 months for codes 324C (11.7% TBTO), 389B (4.9% TBTO), and 351B (9.8% TBTO) all in chloroprene-type rubber matrices was predicted after a 2-y laboratory evaluation.[41]

Various organotin-containing elastomers were subjected to bioassay against *H. trivolvis* in 1965 at the University of Akron. Although loss rates were unknown, 0.1- and 0.01-g specimens destroyed test snails in repeated exposures for six or more months. Berrios-Duran and Ritchie examined the early formulations and confirmed their long-term molluscicidal activity.[25] The material in question, B.F. Goodrich Co. code 443A (the precursor of Nofoul® rubber), is known to release TBTO over a period of less than 9 yr. Consequently, the killing concentrations have been significantly lower than those observed in the evaluation of nonformulated TBTO.[42,43]

Formulations EC-12 and EC-13 (both 30% TBTF) supplied by Environmental Chemicals, Inc. were examined in simple laboratory bioassay against adult *B. glabrata.* Pellet exposures were continuous, with water change prior to each reintroduction of snails. Table 14 depicts LT_{100} values for various size pellets.[44]

Conventional molluscicide treatment results in acute intoxication and rapid mortality of the target. Slow-release treatment established constant low-level intoxication of the habitat, leading to moribundity and eventual mortality through subacute or chronic manifestations. Chronic intoxication does not follow the classical concentration–time dependency, using much less toxicant for a desired mortality.

The nature of the chronic intoxication mechanism has not been elucidated. Haber's rule may be untenable due to organotins inducing multiple responses as discussed by Steinberg,[45] or it may arise from the nonresponsiveness of various physiological protective mechanisms as hypothesized for aquatic plant interactions with herbicidal agents at very low concentrations.[8]

TABLE 14. Evaluation of EC-12 and EC-13 Controlled-Release TBTF Formulations in 1 Liter Water (10 *B. glabrata* per Liter)

Code	Total TBTF Conc. (ppm)	Immersion Time (days)	LT_{100} (days)
EC-12	33.5	0	4
		28	11
		50	10
		62	12
		70	12
		76	13
	17.6	0	7
		18	12
		50	10
		70	7
	7.7	22	13
		39	32
		86	26
	4.7	0	8
		22	30+
EC-13	36.2	0	4
		12	7
		31	8
		36	6
		55	8
		74	7
	18.2	0	5
		12	7
		25	9
		31	8
		50	10
		72	12
		84	9
	8.9	0	8
		12	11
		31	11
		88	26
	6.6	0	15
		22	16
		36	13

TABLE 14 (continued)

Code	Total TBTF Conc. (ppm)	Immersion Time (days)	LT_{100} (days)
		52	35
	6.2	0	12
		22	15
		36	13
		64	35+
	5.9	0	8
		11	8
		24	8
		36	17
	5.8	0	8
		11	10
		11	6
		24	10
		24	6
		37	34
		54	22
		66	38
	5.3	0	7
		11	14
		44	9
		61	35+
	2.7	0	9
		11	13
		24	35+
	2.1	0	12
	1.0	0	14

Conventional methods of assessing relative molluscicidal potency, such as LD_{50}, LC_{90}, and so on, have little meaning in a controlled-release pesticide application situation. In assessing chronic intoxication, mortality is not directly dependent on agent concentration, but rather the interval from treatment time to mortality is proportional to concentration. The vital factor is exposure time, and the most appropriate indices are the LT_{50} and LT_{90}, the time required to achieve a proportional mortality. Decreasing dosage merely increases the time to mortality. Snails will eventually succumb to TBTO or TBTF intoxication even at

10^{-5} or 10^{-6} ppm. Obviously there is a lower limit where the premortality interval is equal to the average lifespan. Unlike conventional molluscicides, where a no-effect concentration exists without regard to exposure duration—about 0.01–0.04 ppm for ethanolamine niclosamide[46] and 0.08 for copper sulfate—ultralow organotin concentrations (10^{-5} ppm and lower) produce irreversible physiological effects inducing mortality when exposure is continuous for 30 or more days.

Snails exposed to very low organotin content over a period of days to weeks will become moribund and eventually succumb. Organotin effects on the snail are dramatic. At levels as low as 0.1 ppb, fecundity is affected and egg laying drastically suppressed. Snail mortality occurs through a chronic intoxication syndrome at 1 ppb and higher. The dosage rate is not controlling with respect to mortality. The end result is the same whether a given population is exposed to 1 or 100 ppb, only the time to a given degree of mortality differs. Death occurs through proteolysis, the destruction of the cell walls within specific connective tissue followed by massive internal hemorrhage. Since specific peptide linkages appear to be involved, as well as possible interruption of amino acid coordination sites, it is unlikely that tolerance can be developed. Indeed it has been noted that the progeny of snails exposed to an LD_{50} of tributyltin fluoride are dramatically less capable of survival than the parent generation.[47] In a recent unpublished study conducted at the University of Akron, differences were shown in the growth rate of progeny of snails exposed to acute and chronic dosages of TBTO. Note in Figure 1 that the first generation of recovering snails exposed to 10 ppm-ta (total active) concentration of TBTO for 3 h exhibit a significantly slower rate of growth and presumably maturation than control snails. However, for the

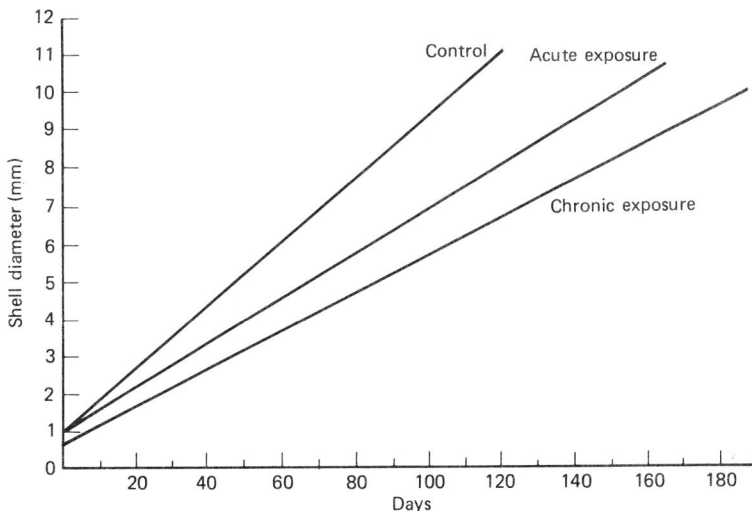

FIGURE 1.   Growth rate of the F_1 generation of *B. glabrata* exposed to TBTO.

progeny of snails exposed to low dosages (0.5 ppm-ta for 36 h) the effect is even more dramatic. This study provides further evidence for the existence of two separate toxic mechanisms in the snail, the "acute" and "chronic" mechanisms resulting from distinct acute and chronic intoxications.

REFERENCES

1. F. Haber, "Zur geschichte des gaskrieges," *Furf Vortrage aus den Jahren 1920-1923*, Julius Springer, Berlin, 1924, pp. 76–92.

2. S. A. Quinn, N. F. Cardarelli, and E. O. Gangstad, *J. Aquatic Plant Manage.*, 15, 74 (1977).

3. S. A. Quinn and N. F. Cardarelli, *Aquatic Herbicide Chronicity Study*, Annual Rep., U.S. Army Corp Eng., Washington, DC, DACW73-72-C-0031, AD903208, unpub. rep., July, 1972.

4. S. Bille, S. Z. Mansdorf, and N. F. Cardarelli, "Development of Slow Release Herbicide Materials for Controlling Aquatic Plants," Final Rep., Dept. Army, Office Chief Engineer DACW73-70-C-0030, Washington, DC, unpub. rep., July 1971.

5. N. F. Cardarelli, *Controlled Release Pesticide Formulations*, CRC Press, Cleveland, OH, 1976, pp. 106–112.

6. N. F. Cardarelli, "The Efficacy, Environmental Impact and Mechanism of Release and Dispersal of Pesticidal Materials Emitted from a Controlled Release Polymeric Dispenser," *Advan. Pesticide Sci.*, 3, 744–753 (1979).

7. G. A. Janes, "Chronicity Phenomenon," in N. F. Cardarelli, Ed., *Proc. Int. Controlled Release Pesticide Symp., Univ. Akron, Akron, OH*, 1974, pp. 14.1–14.13.

8. N. F. Cardarelli and S. V. Kanakkanatt, "A Hypothesis on Chronic Intoxication by Herbicides in Ultralow Concentrations," Am. Chem. Soc. Meet., Univ. Hawaii, Honolulu, Hawaii, June 11–13, 1975 (unpubl. rep.).

9. W. C. Shaw, J. L. Hilton, D. E. Moreland, and L. L. Jansen, "Herbicides in Plants," in *The Nature and Fate of Chemicals Applied to Soils*, Symp. Beltsville, MD, USDA Pub. ARS20-9, April 27–29, 1960, p. 119.

10. R. G. Wilson and H. H. Cheng, *Weed Sci. Am. Meet., Abst.* 77, Feb. 1975.

11. S. S. Q. Hee and R. G. Sutherland, *Weed Sci.*, 21, 115 (1973).

12. J. R. Bauer, R. W. Bovey, and I. Riley, *Weed Sci.*, 22 481 (1974).

13. A. S. Crafts, *The Chemistry and Mode of Action of Herbicides*, Interscience, New York, 1961.

14. J. E, Loeffler and J. Van Overbeek, "Metabolism of Herbicides," in R. White-Stevens, Ed., *Pesticides in the Environment*, Vol. I, Marcel-Dekker, New York, 1971, Part I, p. 237.

15. J. Levitt, *Responses of Plants to Environmental Stress*, Academic, New York, 1972.

16. B. D. Dinman, *Science*, 175, 495 (1972).

17. G. E. Hutchinson, *Proc. Nat. Acad. Sci. U.S.*, 51, 930 (1964).

18. Anon., *Weeds, Trees Turf*, 13(3), 83 (1974).

19. G. A. Janes, "Controlled Release Copper Herbicides," in F. W. Harris, Ed., *Proc. Int.*

Controlled Release Pesticide Symp., Wright State Univ., Dayton, OH, 1975, pp. 326–333.

20. K. E. Walker and N. F. Cardarelli, "Development of Slow Release Copper Sulfate as a Molluscicide," Annual Rep., Int. Copper Res. Assoc., New York, July 1, 1974.

21. N. F. Cardarelli, "Monolithic Polymeric Devices," in A. F. Kydonieus, Ed., *Controlled Release Technologies: Methods, Theory and Application*, Vol. 1, CRC Press, Boca Raton, FL, 1980, pp. 55–72.

22. N. F. Cardarelli, "Monolithic Elastomeric Materials," ibid., pp. 73–128.

23. N. F. Cardarelli and H. F. Neff, Fr. Patent 1506704 (Dec. 19, 1966).

24. N. F. Cardarelli, U.S. Patent 417181 (Dec. 17, 1968).

25. L. A. Berrios-Duran and L. S. Ritchie, *Bull. WHO*, 39 310 (1968).

26. H. S. Hopf and P. H. Goll, "Preliminary Assessment of the Molluscicidal Activity of Biocidal Rubber," Rep. Trop. Pesticide Res. Headquarters, London, Aug. 23, 1968 (Unpub. report).

27. A. Fenwick, "Laboratory Evaluation of Biocidal Pellets 633B for Molluscicide Use," Misc. Rep. 676, Trop. Pesticides Res. Inst., Arusha, Tanzania, 1968 (unpub. report).

28. C. P. Da Souza and E. Paulini, "Laboratory and Field Evaluation of Some Biocidal Rubber Formulations," *WHO Rep., Ser.* PD/MOL/69.9, 1969.

29. C. W. Castleton, "Brazilian Field Trials of MT-1E," *Progr. Rep.* 2, Centro de Pesquisas de Productos Naturias, Rio de Janeiro, Brazil, July 1973 (unpub. report).

30. B. Gilbert et al., *Bull. WHO*, 49, 633 (1973).

31. A. Ross, "Analysis of BioMet SRM Sample Pellets from Brazil," M. & T. Chemicals Inc., Rahway, NJ, correspondence to N.F. Cardarelli, May 8, 1974.

32. A. Fenwick, "Some Observations on Molluscicides in Connection with the Schistosomiasis Pilot Control and Training Project at Misunqwc (Tanzania)," *WHO Rep. Ser.* AFR/BILHARZ/14, Feb. 3, 1969, (Unpub. report).

33. E. S. Upatham, "Preliminary Results on Slow Release TBTO Pellets (MT-1E) against St. Lucian *Biomphalaria glabrata* in a Marsh and a Ravine," *Proc. Int. Controlled Release Pesticide Symp., Univ. Akron, Akron, OH*, 1976.

34. A. Mansouri, "Brief Summary of a Study of MT-1E Slow Release Molluscicide," *Proc. Int. Controlled Release Pesticide Symp., Univ. Akron, Akron, OH*, 1976, p. 5.52.

35. A. Santos, "Field Trials with CBL-9B and MT-1E in Leyte," *Res. Note*, Schistosomiasis Control and Research Service, Dept. Health, Manila, Philippines, Sept. 30, 1976.

36. J. Elsa and V. Paynter, *AMA Arch. Indust. Health*, 18, 244 (1958).

37. C. J. Shiff, "Focal Control of Schistosome-Bearing Snails," in T. C. Cheng, Ed., *Molluscicides in Schistosomiasis Control*, Academic, New York, 1974, pp. 241–247.

38. A. Santos, "Summary of Results of Laboratory Screening of Compound 443A against *O. quadrsi*," *Proc. Int. Controlled Release Pesticide Symp., Univ. Akron, Akron, OH*, 1976, p. 5.62.

39. E. A. Malek, "Bilharziasis Control in Pump Schemes Near Khartoum, Sudan and an Evaluation of the Efficacy of Chemical and Mechanical Barriers," *Bull. WHO*, 27, 41 (1962).

40. H. S. Hopf and P. H. Goll, *WHO Rep. Ser.*, PD/MOL/69.4, 1969.

41. H. S. Hopf and P. H. Goll, "Biocidal Rubber: A Two Year Assessment of the Organotin-Containing Formulations," *Rep. Trop. Pesticides Res. Headquarters*, London, 1969.

42. L. S. Ritchie and E. A. Malek, "Molluscicides: Status of Their Evaluation Formulations and Methods of Application," *WHO Rep. Ser.*, PD/MOL/69.1, 1969.

43. L. S. Ritchie et al., Molluscicidal Time-Concentration Relationships of Organotin Compounds," *Bull. WHO*, **31**, 147 (1964).

44. N. F. Cardarelli, "Evaluation of Environmental Chemicals Inc. Controlled Release Materials," Rep. (unpub.), Creative Biology Laboratory, Barberton, OH, Sept. 1976.

45. M. Steinberg, "The Toxicology of Controlled Release," in *Proc. Int. Controlled Release Pesticide Symp.*, Nat. Bur. Standards, Gaithersburg, MD, 1978, pp. 2.42–2.52.

46. M. M. Ishak and A. M. Mohamed, "Effects of Sublethal Doses of Copper Sulfate and Bayluscide on Survival and Oxygen Consumption of the Snail *Biomphalaria alexandria,*" *Hydrobiol.*, 47, 499–512 (1975).

47. N. F. Cardarelli, "Laboratory and Field Evaluations of Controlled Release Molluscicides and Schistolarvicides," Annual Rep. (unpub.), Edna McConnell Clark Foundation, 276-0091, New York, 1977.

Index